MathPassion

Mamdouh Hamouda

To the soul of my father

CreateSpace, Charleston SC

Contents

PREFACE

Those who have passion with mathematics feel great joy as they indulge in solving a challenging mathematical riddle, when they are busy reading about a mathematical problem that has been puzzling mathematicians of antiquity or reading an elegant proof of a theorem that reveals admirable insight.

This feel of "*mathing*" is – in a way – similar to the feeling when you listen to good music or watch a beautiful piece of art.

However, we also know that there are many other people who have not been exposed sufficiently to math practicing and who would be equally enjoying to sail in its sea if they have the opportunity to get more familiar with its treasures

You do not have to be a mathematician to appreciate the reasoning of a shrewd mathematical proof or enjoy brain stimulation brought by a mathematical riddle.

In this book, I am trying to explore the elements of beauty in mathematics in a journey through classical and modern mathematical problems and accomplishments.

Topics discussed are those that would intrigue most of us and not those typically included in standard school syllabuses.

Because proofs – in particular - constitute beautiful and brain stimulating facets of mathematics; I tried not to miss the reasoning and logic underlying them whenever a conjecture, theorem or formula is discussed.

While doing so I focused – in as much as was possible - on graphical presentations and used proof techniques such as:

- Proof by rearrangement

- Proof by contradiction

- Proof by Mathematical Induction

- Proof by performing a Thought Experiment

Chapters are not directly interrelated, so the reader would feel free to cruise directly to the topic that appeals to him more.

There is a Chapter Quiz followed by a discussion and solution at the end of each chapter (except chapter ten). This is not intended to test the reader's knowledge but to offer complementary relevant information in a rather stimulating and interactive form

Chapter 1:

Pythagorean Triples

**Figure 1-1: Sculpture of
Pythagoras at Pitagora
Museum, Crotone Italy**

The Pythagorean Theorem is probably the most famous
mathematical statement and definitely one the of most important
theorems. It is so named after the great Greek philosopher, scientist
and mathematician "**Pythagoras**" (572-479 BC).

Pythagoras was born in Samos of Greece and spent much time in
Egypt – as several of Greek mathematicians did – and in Babylonia.
At his forties, he settled in Crotona which is a Greek town in southern
Italy where he formed a group of followers known as the
Pythagoreans. That was a religious, spiritual and mathematical
group. They followed a code of secrecy which portrayed a historical
image for Pythagoras as a mystic figure who gained a great respect
by his followers *(Katz 2009)*.

Pythagoras Theorem states that the sum of the square of shorter
legs in a right angled triangle equals the square of the hypotenuse,
that is if "**a**", "**b**" are the lengths of legs forming a right angle in a
triangle and "**c**" is the length of the hypotenuse then:

$$a^2 + b^2 = c^2$$

Pythagoras Theorem (also known as the Pythagorean Theorem) was thought by its creator as a purely geometric theorem, as necessitated by application requirements at the time.

Now it is also viewed to support algebraic applications as well, and in this context I shall focus on its algebraic interface with Number Theory – the branch of mathematics that deals with whole numbers.

It is established that the Pythagorean Theorem was known several centuries before Pythagoras, but he was probably the first one to prove it. It was subsequently proven by numerous mathematicians who generated hundreds of proofs. Probably the most authentic proof brought up for the theorem was that made by Euclid in Book 1 of his encyclopedic set of books "*The Elements*".

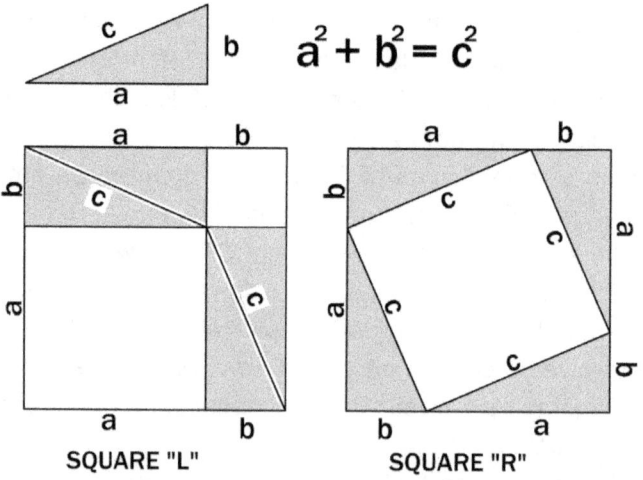

Figure 1-2: Two squares congruent at outline but arranged differently inside

Another proof, controversially believed to have been made by Pythagoras or by a Chinese mathematician is known as "proof by rearrangement". It is an amazingly simple graphical proof that requires the least possible textual argument

The two large squares "**L**" and "**R**" shown in figure 1-2 are congruent at the outline as they have the same side length of "**a + b**", where "**a**" and "**b**" are the length of the shorter legs forming the right angle of the shaded triangle shown in the figure.
The area of square "**L**" therefore equals the area of square "**R**" which numerically equals: $(a + b)^2$, for $(a + b)$ being the side length of the outline square "**L**" or "**R**".

You will notice that our shaded right angled triangle having side lengths of "**a, b & c**", is copied four times in each of the two large squares "**L**" and "**R**".
Now if we remove these four triangle from the two squares; the remaining part of the shapes of "**L**" and "**R**" will be as shown in figure 1-3, and since the two large squares "**L**" and "**R**" are congruent at outline we should expect that the remaining shapes in the figure (after trimming off the 4 triangles) will also be of equal areas

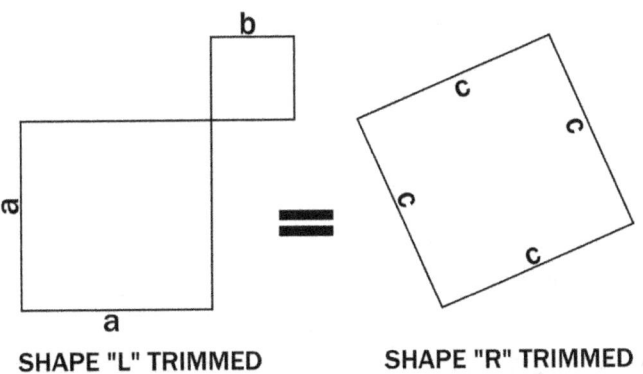

SHAPE "L" TRIMMED   **SHAPE "R" TRIMMED**

Figure 1-3: The two squares trimmed off by
removing the 4 identical triangles (a, b & c)

The remaining shapes are nothing but the two squares at the triangle's legs: (having the areas a^2 and b^2) on left side and the square c^2 at the hypotenuse on right side, hence the theorem has been graphically proven.
The measures of the triangle sides **a**, **b** and **c** in the general

form are real numbers.

For example if $a = 2$ and $b = 3$, $c^2 = a^2 + b^2 = 2^2 + 3^2 = 4 + 9 = 13$, hence $c = \sqrt{13} = 3.6055...$

Obviously, this is not a whole number nor the fraction it contains is rational, and additional digits at the right side of the decimal point will appear endlessly towards a better approximation for the square root of the number **13**.

In the special case of having **a**, **b** & **c** measures represented integrally i.e. as whole numbers, we call these three whole numbers a "*Pythagorean Triple*".

A **Pythagorean Triple** is therefore a set of three whole numbers **a**, **b** & **c** that satisfy the equation: $a^2 + b^2 = c^2$

A well-known example of the Pythagorean Triple is the triangle in which the measures of the two legs and the hypotenuse are **3**, **4** & **5** respectively as it is obvious that $3^2 + 4^2 = 5^2$ (or **9 + 16 = 25**).

The (**3, 4, 5**) triangle had been known nearly a thousand years before Pythagoras was born. It is known as the Egyptian Triangle because ancient Egyptians used to have a twelve yards rope knotted at equally spaced 11 intermediate points as a tool to set out the right angle at temples and pyramids as can be seen in figure 1-5

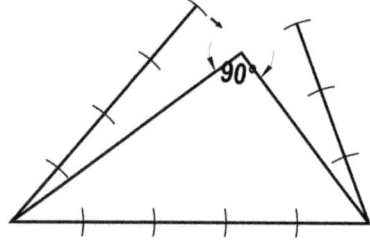

Figure 1-4: The great pyramid

Figure 1-5 : The 12 yards knotted rope used to construct a right angle

There is an infinite number of other Pythagorean Triples. A few examples are (**5,12,13**), (**7,24,25**), (**8, 15, 17**), (**9,40,41**) .... and so many others.

Furthermore, any of these triples can be scaled up by multiplying it by any integer greater than one. For example if our renowned Pythagorean Triple (**3,4,5**) is factored by **2**, **3**, **4** and **5**; we get the following Pythagorean Triples respectively: (**6,8,10**), (**9,12,15**) , (**12,16,20**) and (**15, 20, 25**).

While these Pythagorean Triples satisfy the equation $a^2 + b^2 = c^2$ integrally (i.e. with only whole numbers used), this kind of triples is called "**non-primitive**".

A "**Primitive Pythagorean Triple**" is that in which **a**, **b** and **c** are relatively co-prime. In other words "Primitive Pythagorean Triple" is a Pythagorean Triple in which **a**, **b** and **c** cannot be reduced by division by any common factor. So while (**3,4,5**) is indeed a primitive Pythagorean Triple, triples such as (**6,8,10**), (**9,12,15**) or (**12,16,20**) are non-primitive.

I shall focus here on the means of generating primitive Pythagorean Triples. There are several ways to do that and I shall start by introducing a couple of methods which are simple and easy to use yet capable of sorting out a great deal – but indeed not all – of Pythagorean Triples.

For the sake of discussion, let us call the first method to be presented here "***The Odd Numbers Procedure***"

To apply the *Odd Numbers Procedure* we shall start by selecting any odd number, say "**3**".

If you square it you will also get another odd number (**9**). Now split the **9** into two numbers having a difference of one, so in this example you will get "**4**" and "**5**".

The odd number you initially selected (**3**) along with two numbers that sum up to the square of selected number (**4,5**) will form the primitive Pythagorean Triple (**3, 4, 5**).

Continue the same process using other odd numbers and you will get a primitive Pythagorean Triples for every odd number.

This infinite number of triples equals half the number of all natural numbers. I will be showing in the tabulation in figure 1-6 only a few examples for Triples generated by this method, and the reader may extrapolate as many Pythagorean Triples as he/ she wishes, by following the same procedure.

Odd number (K)	K^2	K2 is split into two numbers having a difference of "1"		The primitive Pythagorean triple generated
5	25	12	13	(5,12,13)
7	49	24	25	(7,24,25)
9	81	40	41	(9,40,41)
11	121	60	61	(11,60,61)
13	169	84	85	(13,84,85)
15	225	112	113	(15,112,113)
17	289	144	145	(17,144,145)
19	361	180	181	(19,180,181)

Figure 1-6: Generating Pythagorean Triples starting with odd numbers

You may wish to interrogate the validity of the Pythagorean Triples generated from odd numbers shown in figure 1-6 with regard to its conformance to the Pythagorean Theorem, and I owe the reader a proof of such conformity.

The Pythagorean Triple generated from odd numbers has the following composition pttern:
a: Shorter leg is the odd number **K**
b: Longer leg $\qquad$ $(K^2 - 1)/2$
c: The hypotenuse $\qquad$ $(K^2 + 1)/2$

$$a^2 + b^2 = K^2 + (K^2 - 1)^2/4 = K^2 + (K^4 - 2K^2 + 1)/4 = (K^4 + 2K^2 + 1)/4 = (K^2 + 1)^2/4 \quad \dots\dots\dots\dots\dots\dots\dots (1)$$

$$c^2 = (K^4 + 2K^2 + 1)/4 = (K^2 + 1)^2/4 \dots\dots\dots\dots\dots\dots (2)$$

From (1) and (2): $a^2 + b^2 = c^2$, hence the "Odd Numbers" method correctly generates Pythagorean Triples.

But why should we start with odd numbers? Can we generate Pythagorean Triples starting with even numbers?

Indeed, we can, and this will lead us to the second method. Let us call it "***The Even Numbers Procedure***".

Take an even number **a** in the form of **2 j**.
So if **j =3, a** = 2 * 3 = **6** will be the shorter leg of the triangle.

The longer leg in this example will be: **b = j^2 - 1** = **3^2** - **1** = **8**, and the hypotenuse **c** will be: **j^2 + 1** = **3^2** - **1** = 9 + 1 = **10**

The Pythagorean Triple formed by this procedure is (**6**, **8**, **10**)
You will notice that this triple is similar to the (**3**, **4**, **5**) triple with a multiplier of 2 introduced, hence it is a non-primitive triple because it can be reduced to the (3,4,5) triple if divided by 2.

Primitiveness may be examined in the tabulation shown in figure 1-7 which presents a few examples of Pythagorean Triples generated starting with Even Numbers method.
We can directly spot that half the triples generated by this method are non-primitive as they can be reduced by halving. However, if you start with an even number which is also divisible by 4, the outcome will be a primitive triple. Why is that?
Starting with an even number in the form of **2 j** which is not divisible by 4 means that j is an odd number and so will be **j^2**. Now since **j^2** is odd then the other two numbers forming the triple (**j^2 - 1** and (**j^2 + 1**) will obviously be even. Since the three numbers forming the triple are even as such then we are having a non-primitive triple.

Even number (2j)	j^2	j2 - 1	j2 + 1	The Pythagorean triple generated	Comment
6	9	8	10	(6,8,10)	Non-primitive
8	16	15	17	(8,15,17)	
10	25	24	26	(10,24,26)	Non-primitive
12	36	35	37	(12,35,37)	
14	49	48	50	(14,48,50)	Non-primitive
16	64	63	65	(16,63,65)	
18	81	80	82	(18,80,82)	Non-primitive
20	100	99	101	(20,99,101)	

Figure 1-7: Generating Pythagorean Triples starting with even numbers

In conclusion, if we seek only primitive triples then we must extract from the table in figure 1-7 only the triples corresponding to "a" which is divisible by 4 (those which are not highlighted). Highlighted triples should be excluded.

Now we got to prove that triples generated using the even number procedure are Pythagorean.

The Composition Pattern for this triple is:

$$a \ = 2\,j, \qquad a^2 = 4\,j^2$$
$$b \ = j^2 - 1 \ , \qquad b^2 = (j^2 - 1)^2 \ = j^4 - 2\,j^2 + 1$$
$$c \ = j^2 + 1 \ , \qquad c^2 = (j^2 + 1)^2 \ = j^4 + 2\,j^2 + 1$$

$a^2 + b^2 \ = \ j^4 + 2\,j^2 + 1 \ = \ c^2$, hence the Even Numbers method correctly generates Pythagorean Triples.

Triangles having the Pythagorean proportions of the triples provided by the "Even Numbers" procedure can be generated graphically on an orthogonal grid of two rows of squares, by joining the grid nodes with diagonal line segments as shown in figure 1-8

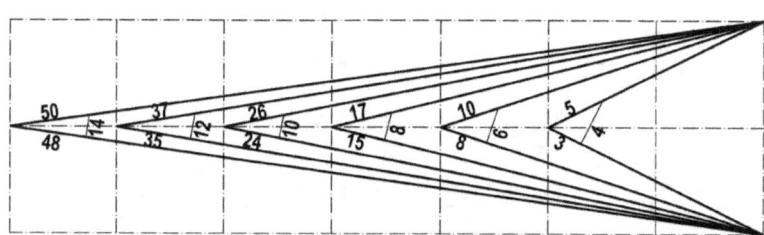

Figure 1-8: A graphical method to construct triangles having the Pythagorean proportions generated by the Even Numbers method

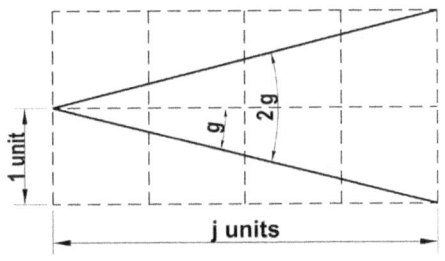

Figure 1-9: The proportions of the tangent of a double angle

j units

To prove the analogy with the Even Numbers procedure, consider the two lines symmetrically connecting nodes of the orthogonal squared grid in figure 1-9.

The tangent of twice the angle may be calculated using the following trigonometric formula:

$$\tan(2g) = 2\tan(g) / [1 - (\tan^2(g)]$$

$$\tan g = 1/j, \quad \text{hence } \tan(2g) = \frac{2\left(\frac{1}{j}\right)}{1 - \frac{1}{j^2}} = \frac{2j}{j^2 - 1}$$

This slope is attained if we have the length of the right angle legs: $a = (2\,j)$ and $b = (j^2 - 1)$ which are just the same expressions we had in our discussion to prove the validity of the Even Numbers procedure. To refine the process by merging the Odd and Even Numbers procedures in one combined method that generates primitive Pythagorean Triples we may proceed as follows:

- Start by picking a number "**a**" to be the first number in the triple

- If "**a**" is an odd number; square it then split the resulting square into two numbers having a difference of (1) between them. These will be the other two numbers in the triple:

$$b = (a^2 - 1)/2 \qquad \text{and} \qquad c = (a^2 + 1)/2$$

- If "**a**" is an even number; halve it and square this half then subtract and add (1) from/ to it respectively to get the other two numbers in the triple:

$$b = (a/2)^2 - 1 \qquad \text{and} \qquad c = (a/2)^2 + 1$$

However, if only primitive triples are to be generated, we must exclude even numbers which are not divisible by 4

A flowchart for the process is shown in figure 1-10 while figure 1-11 lists the first 10 primitive Pythagorean Triples generated by the combined procedure.

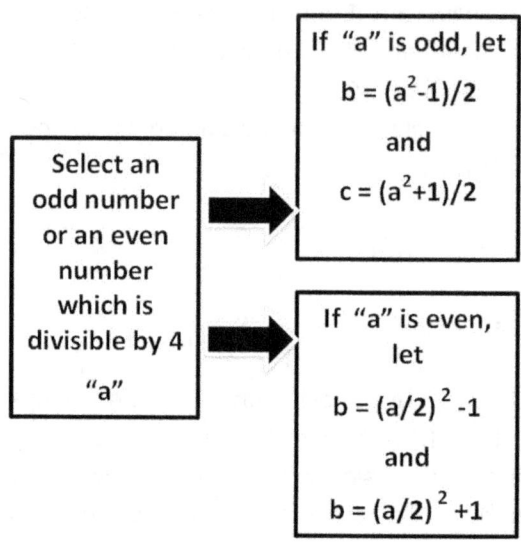

Figure 1-10:

The Combined Procedure to generate primitive Pythagorean Triples

Figure 1-11:

A list of the first 10 primitive Pythagorean Triples generated by the Combined Procedure

s/n	a	b	c
1	3	4	5
2	5	12	13
3	7	24	25
4	8	15	17
5	9	40	41
6	11	60	61
7	12	35	37
8	13	84	85
9	15	112	113
10	16	63	65

The combined procedure is a practical and efficient tool to generate primitive Pythagorean Triples.
First 10 triples generated by the combined procedure are shown in figure 1.11. A lot more can be generated using a standard spreadsheet based software

The procedure generates a primitive triple for every odd value of "a" and for the half of even values that divide 4.

The number of primitive triples generated by the procedure amounts to 75% of the whole of the natural numbers.
But does the procedure conclusively generate all possible primitive Pythagorean Triples? The answer is No.

For example, triples like (**33, 56, 65**) or (**204, 253, 325**) are indeed Pythagorean – and the reader is invited to examine that by checking whether the sum of squares of the first two numbers equals the square of the third.
Yet these two triples are not generated through the combined procedure, and it seems that we have to find another method that offers complete conclusiveness.

We do have this method!

It is known as the **Euclid's Formula**. This is a fundamental formula for the generation of Pythagorean Triples.

If you have a pair of positive integers (whole numbers) **m** and **n** where m > n, then following Composition Pattern would form a Pythagorean Triple:

2 m.n   (Obviously this is an even integer)

$m^2 - n^2$ (This will be odd only if either **m** or **n** is odd and the other is even)

$m^2 + n^2$ (this is the hypotenuse. It will be odd only if either **m** or **n** is odd and the other is even)

It is important for the generation of primitive triples to observe the following hints:

- Only one of **m** & **n** should be an odd number with the other being even. If both **m** and **n** are odd; **2.m.n**, ($m^2 - n^2$) and ($m^2 + n^2$) will all become even hence there will be a common factor of 2. Same will be the case if both **m** & **n** are even

- **m** and **n** should be relatively prime, that is they should not have a common factor.

Let us examine the compliance of this formation with the Pythagorean Theorem.

	a	b	c
Sides/hypotenuse	2 m.n	$(m^2 - n^2)$	$(m^2 + n^2)$
(Sides/hypotenuse)^2	$4 m^2. n^2$	$(m^4 - 2m^2n^2 + n^4)$	$(m^4 + 2m^2n^2 + n^4)$

From above tabulation: $a^2 + b^2 = m^4 + 2 m^2.n^2 + n^4 = c^2$, hence Euclid's Formula - just validated - does generate Pythagorean Triples.

The two triples (**33, 6, 65**) and (**204, 253, 325**) given in the example above have been generated by Euclid's formula as explained below:

(**33, 56, 65**) resulted from: **m= 7   and n =4**
(**204, 253, 325**) resulted from: **m= 17 and n =6**

Figure 1-12 lists other 20 primitive Pythagorean Triples generated by Euclid's formula.
Let us now wrap up the search for Pythagorean Triples and agree on a methodology to generate them systematically.
Consider the (**2 mn,   m^2 - n^2,   m^2 + n^2**) Composition Pattern of **Euclid**'s formula. For n=1, the Euclid's composition pattern will turn out to take the shape: (**2 m, m^2 - 1, m^2 + 1**)
You will notice that this nothing but the Composition Pattern of the **Even Numbers** procedure. We simply replaced "**j**" in the just discussed **Even Number** procedure by "**m**"

Furthermore, if you divide the above terms of Euclid's composition pattern by **2** you will get:

$$m, \quad \frac{m^2-1}{2}, \quad \frac{m^2+1}{2}$$

And sure enough you will notice that this latter composition is nothing but the composition pattern of the Odd Numbers procedure; with "**k**" replaced by "**m**".

What does this mean? It simply means that Euclid's formula is the general tool to generate Pythagorean Triples, and that the Combined Procedure – which is the merge of Odd and Even numbers procedures – represents the special case of the Euclid's formula in which n =1. However such a special case caters for nearly 70% of the number of all primitive Pythagorean Triples. Yet this simplified procedure is far easier to handle for its being dependent on a single variable.

s/n	m	n	m*n	a and b		c	(a,b,c),
	A LIST OF 20 PRIMITIVE PYTHAGORIAN TRIPLES						
	EXCLUSIVELY GENERATED BY EUCLID'S FORMULA						
1	2	5	10	20	21	29	(20, 21, 29),
2	2	7	14	28	45	53	(28, 45, 53),
3	2	9	18	36	77	85	(36, 77, 85),
4	2	11	22	44	117	125	(44, 117, 125),
5	2	13	26	52	165	173	(52, 165, 173),
6	4	7	28	33	56	65	(33, 56, 65),
7	3	10	30	60	91	109	(60, 91, 109),
8	2	15	30	60	221	229	(60, 221, 229),
9	2	17	34	68	285	293	(68, 285, 293),
10	10	13	130	69	260	269	(69, 260, 269),
11	11	14	154	75	308	317	(75, 308, 317),
12	2	19	38	76	357	365	(76, 357, 365),
13	3	14	42	84	187	205	(84, 187, 205),
14	2	21	42	84	437	445	(84, 437, 445),
15	6	11	66	85	132	157	(85, 132, 157),
16	7	10	70	51	140	149	(51, 140, 149),
17	2	23	46	92	525	533	(92, 525, 533),
18	2	25	50	100	621	629	(100, 621, 629),
19	2	27	54	108	725	733	(108, 725, 733),
20	9	14	126	115	252	277	(115, 252, 277),

Figure 1-12: A list of 20 primitive Pythagorean triples generated by Euclid's formula

Having gone thus far in analyzing the Pythagorean Triples let us extend our search further to explore the realm of three dimensions.

The intent is not to raise the exponent of a, b, and c to 3. This is a taboo which I won't dream about it because it hits the core of Fermat's Last Theorem, about which I shall give a hint shortly later. What I mean is that I shall be searching for an integral solution for the equation $A^2 + B^2 + C^2 = D^2$
Where such a relationship would be encountered?

Consider the 3D body drawn in figure 1-13. This right angled body is known as the cuboid. It is a cube that has been stretched in different proportions in the three directions of space.

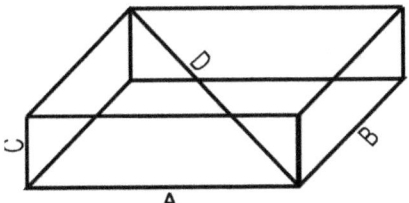

Figure 1-13: A cuboid of the dimensions A, B, C and a hypotenuse D

The question is whether we can have an integral solution for the equation

$A^2 + B^2 + C^2 = D^2$; in the same way we had with the Pythagorean Triples?

Let us call the variables **A**, **B**, **C** and **D** that satisfy the above equation the "**Pythagorean Quadruple**".
The answer is yes we can. I shall not elaborate on that matter much but shall mention a few formulae that generate Pythagorean Quadruples.

One simple way to generate the **Pythagorean Quadruple** is to take any whole number **k**, the subsequent number **k+1** and the multiple of these two numbers **k*(k + 1)** as **A**, **B** and **C** respectively.
The hypotenuse **D** in this case will be: **k*(k + 1) + 1:**

Example:

2, (2+1) = **3**, (2 *3) = **6** and (6+1) = **7**)

Indeed: $2^2 + 3^2 + 6^2 = 4 + 9 + 36 = 49 = 7^2$

In general, sum of the square of the three sides of the cuboid: **A**, **(A+1)** and **A(A+1)** is

$A^2 + (A + 1)^2 + A^2(A + 1)^2 = A^2 + (A^2 + 2A + 1) + A^2(A^2 + 2A + 1)$
$= A^4 + 2A^3 + 3A^2 + 2A + 1$ .... (1)

And the square of the hypotenuse **[A (A + 1) + 1]** is

$$A (A + 1) + 1\}^2 = (A^2 + A + 1)^2 = A^4 + 2(A^3 + A^2) + (A^2 + 2A + 1)$$

$$= A^4 + 2A^3 + 3A^2 + 2A + 1 \qquad ......(2)$$

The right hand side in equation (**1**) is identical to the right hand side in equation (**2**) hence the left hand side in equation (**1**) equals the left hand side in equation (**2**):

$$A^2 + (A + 1)^2 + (A* (A + 1))^2 = (A*(A + 1) + 1)^2$$

I must however hint that this is not the only procedure that can generate Pythagorean Quadruples.

There are other several formulae that do the same thing.
Try this one for example:

Take any number **K**, its double **2 K** and $(K^2 - 5)/2$ to represent **A**, **B** and **C** respectively. The hypotenuse **D** in this case will be $D = (K^2 + 5)/2$

The proof:

$$A^2 + B^2 + C^2 = [K^2] + [4 K^2] + [(K^2 - 5)/2]^2$$
$$= (K^2 - 5)^2/ 4 + 5 K^2$$

$$.........(3)$$

$$D^2 = (K^2 + 5)^2/ 4 = (K^2 - 5)^2/ 4 + 4*(K^2*5)/ 4$$
$$=(K^2 - 5)^2/ 4 + 5 K^2 \qquad ....(4)$$

The right hand side in equation (3) is identical to the right hand side in equation (4) hence the left hand side in equation (3) equals the left hand side in equation (4)

$$A^2 * B^2 + C^2 = D^2 \text{ or}$$

$$(A)^2 + (2 A)^2 + (\frac{(A^2 - 5)}{2})^2 = (\frac{(A^2 + 5)}{2})^2$$

While discussing the Pythagorean triples we should not miss Fermat's Last Theorem. A riddle which stimulated the brains of medieval as well as contemporary mathematicians proceeds as follows:

can we have an integral solution for the equation:
$A^n + B^n = C^n$ *for n>2*

A note was written in 1637 by the French mathematician **Pierre de Fermat** (1601 – 1665) on a margin of Diophantus book Mathematica that says:

"*I discovered a proof that* $(A^n + B^n = C^n)$ *has no integral solution for n > 2, but it is too long to fit in this small margin*".

Fermat's proof was never found and the riddle had haunted the world mathematicians for more than 3.5 centuries until it was worked out and proved by Andrew Wiles and released in 1994.

The Chapter Quiz

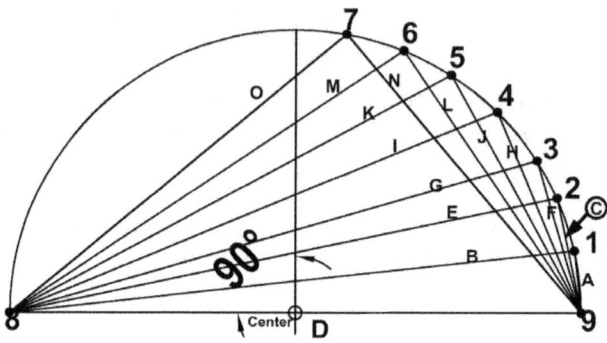

Figure 1-14: The Chapter Quiz

The triangles shown in figure 1-14 above have a common base "D" which is the diameter of the semicircle that circumscribes these triangles. The vertices of these triangles (points 1 to 7) fall on a single quadrant of the circle as shown. Given that the base length and the position of vertices have been carefully selected such that:

- All the sides and common base of the triangles (**A** through **O**) have integral measure (whole number of measuring units)
- The measure of "**D**" is the least possible

Find the measure of A through O.

Discussion and solution of the Chapter Quiz

The reader may notice that while the quiz is almost free of numbers – except the reference to seven vertices – it requires the determination of 15 numbers (A through O) for solution.

Another significant remark is that the seven triangles – being inscribed in a semicircle, with its diameter as a common base – should have a right angle at the circumferential vertex. Why?. Look at the triangle 1-2-3 in figure 1-15 which is also circumscribed by a semicircle. The triangle is split into the two isosceles triangles which meet at the circle centre. They are obviously isosceles because two sides in each triangle are formed by the circle radii. The base interior angles in each of the two triangles will therefore be congruent as indicated in the figure

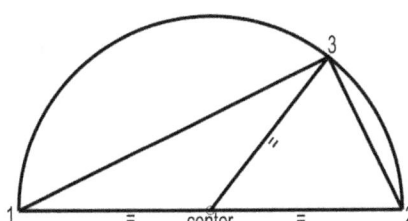

Figure 1-15

Sum of the angles in the larger triangle (1-2-3) = **u + v + u + v = 2(u+v) = 180°** as in any triangle. This means that the measure of angle "**3**" = **u + v = 90°**

Combining this finding with the given condition that the measures of the sides of all triangles are integral, we realize beyond any doubt that it is a question of Pythagorean Triples.

Let us start – as an exploratory effort – by listing the first thirty Pythagorean Triples, using the Combined Procedure. The outcome is the list of 30 triples shown in figure 1-16.

We shall then look at the "**c**" column and recall that our goal is to find the shortest common hypotenuse with an integral measure for seven Pythagorean triangles.

Since the "**c**" values differ, we expect that a certain positive multiplier should be introduced to the carefully selected Triples such that the factored "**c**" will become identical for all cases.

The first three values of "**c**" in the table of figure 1-16 (**5**, **13** and **25**) obviously constitute the least measures for "**c**". Let us practice the optimization technique on these three values.

As discussed, we should introduce a multiplier to each of these three values such that we get one measure for the hypotenuse in the three triples as shown in figure 1-17.

s/n	a	b	c
1	3	4	5
2	5	12	13
3	7	24	25
4	8	15	17
5	9	40	41
6	11	60	61
7	12	35	37
8	13	84	85
9	15	112	113
10	16	63	65
11	17	144	145
12	19	180	181
13	20	99	101
14	21	220	221
15	23	264	265
16	24	143	145
17	25	312	313
18	27	364	365
19	28	195	197
20	29	420	421
21	31	480	481
22	32	255	257
23	33	544	545
24	35	612	613
25	36	323	325
26	37	684	685
27	39	760	761
28	40	399	401
29	41	840	841
30	43	924	925

Figure 1-16: First 30 primitive triples generated by the Combined Procedure

SELECTED TRIPLE				FACTORED TRIPLE		
a	b	c	Multi plicati	a	b	c
3	4	5	65	195	260	325
5	12	13	25	125	300	325
7	24	25	13	91	312	325

Figure 1-17

The process we have just performed now is known as finding the "Least Common Multiple". You will find further elaboration on the Least Common Multiple in chapter 5.

In the problem we are trying to solve, we have a common hypotenuse which is the circle's diameter.

The three non-primitive Pythagorean Triples at the right-side part of the tabulation of figure 1-17 have a common hypotenuse which equals 325, and it is indeed the smallest possible common hypotenuse for three Triples.

But we still have a bonus hidden in the table in figure 1-16 !

With little more attention, we can spot other two triples that can be added to our catch:

- The triple #10, **a=16 b=63   c=65**, which simply needs to be factored by **5** to join the club:
  a= 80 b = 315 c = 325 and indeed, this latter "**325**" is what should qualifies the triple to be added to our selection.

- The triple **#25, a=36 b = 323 c = 325**. This needs no further factoring or seeking least common multiples since "**c**" already equals **325**,

The five triples having the least common hypotenuse that have been depicted so far are tabulated in figure 1-18.

You know that we are still short of two triples. Suppose we decide to pick the next two smaller value for "**c**" from the last column in the table of figure 1-16, we would be adding **c = 17** (**#4**) and **c = 37** (**#7**).

Since both are prime numbers, we would be shooting the common hypotenuse "c" high by incorporating their product in the least common multiple's circle.

	a	b	c
1	36	323	325
2	80	315	325
3	91	312	325
4	125	300	325
5	195	260	325

Figure 1-18

Accordingly, common hypotenuse "**D**" – currently equalling 325 - will become:**325** * **17** * **37** = **204,425** which is a huge rise (remember that the measure of "**D**" should be the least possible)

Do we have any other option? You may recall that while the Combined Procedure is quick and more manageable for its dealing with a single variable, it does not assume complete generality and coverage as that attained by the more comprehensive method based on Euclid's formula.

Let us therefore seek the remaining two triples through the algorithm of Euclid's formula

In order to minimize non-primitive redundancy, only one of the two variables (m,n) is taken as even number while the other one is odd.

The tabulation in figure 1-19 lists the primitive Pythagorean Triples for **m** and **n** not exceeding 30.

n =	2	4	6	8	10	12	14	16	18	20	22	24	26	28	30
m															
1	3	8	12	16	20	24	28	32	36	40	44	48	52	56	60
	4	15	35	63	99	143	195	255	323	399	483	575	675	783	899
	5	17	37	65	101	145	197	257	325	401	485	577	677	785	901
3	5	7	27	48	60	72	84	96	108	120	132	144	156	168	180
	12	24	36	55	91	135	187	247	315	391	475	567	667	775	891
	13	25	45	73	109	153	205	265	333	409	493	585	685	793	909
5	20	9	11	39	75	119	140	160	180	200	220	240	260	280	300
	21	40	60	80	100	120	171	231	299	375	459	551	651	759	875
	29	41	61	89	125	169	221	281	349	425	509	601	701	809	925
7	28	33	13	15	51	95	147	207	252	280	308	336	364	392	420
	45	56	84	112	140	168	196	224	275	351	435	527	627	735	851
	53	65	85	113	149	193	245	305	373	449	533	625	725	833	949
9	36	65	45	17	19	63	115	175	243	319	396	432	468	504	540
	77	72	108	144	180	216	252	288	324	360	403	495	595	703	819
	85	97	117	145	181	225	277	337	405	481	565	657	757	865	981
11	44	88	85	57	21	23	75	135	203	279	363	455	555	616	660
	117	105	132	176	220	264	308	352	396	440	484	528	572	663	779
	125	137	157	185	221	265	317	377	445	521	605	697	797	905	1021
13	52	104	133	105	69	25	27	87	155	231	315	407	507	615	731
	165	153	156	208	260	312	364	416	468	520	572	624	676	728	780
	173	185	205	233	269	313	365	425	493	569	653	745	845	953	1069
15	60	120	180	161	125	81	29	31	99	175	259	351	451	559	675
	221	209	189	240	300	360	420	480	540	600	660	720	780	840	900
	229	241	261	289	325	369	421	481	549	625	709	801	901	1009	1125
17	68	136	204	225	189	145	93	33	35	111	195	287	387	495	611
	285	273	253	272	340	408	476	544	612	680	748	816	884	952	1020
	293	305	325	353	389	433	485	545	613	689	773	865	965	1073	1189
19	76	152	228	297	261	217	165	105	37	39	123	215	315	423	539
	357	345	325	304	380	456	532	608	684	760	836	912	988	1064	1140
	365	377	397	425	461	505	557	617	685	761	845	937	1037	1145	1261
21	84	168	252	336	341	297	245	185	117	41	43	135	235	343	455
	437	425	405	377	420	504	588	672	756	840	924	1008	1092	1176	1260
	445	457	477	505	541	585	637	697	765	841	925	1017	1117	1225	1341
23	92	184	276	368	429	385	333	273	205	129	45	47	147	255	371
	525	513	493	465	460	552	644	736	828	920	1012	1104	1196	1288	1380
	533	545	565	593	629	673	725	785	853	929	1013	1105	1205	1313	1429
25	100	200	300	400	500	481	429	369	301	225	141	49	51	159	275
	621	609	589	561	525	600	700	800	900	1000	1100	1200	1300	1400	1500
	629	641	661	689	725	769	821	881	949	1025	1109	1201	1301	1409	1525
27	108	216	324	432	540	585	533	473	405	329	245	153	53	55	171
	725	713	693	665	629	648	756	864	972	1080	1188	1296	1404	1512	1620
	733	745	765	793	829	873	925	985	1053	1129	1213	1305	1405	1513	1629
29	116	232	348	464	580	696	645	585	517	441	357	265	165	57	59
	837	825	805	777	741	697	812	928	1044	1160	1276	1392	1508	1624	1740
	845	857	877	905	941	985	1037	1097	1165	1241	1325	1417	1517	1625	1741
31	124	248	372	496	620	744	765	705	637	561	477	385	285	177	61
	957	945	925	897	861	817	868	992	1116	1240	1364	1488	1612	1736	1860
	965	977	997	1025	1061	1105	1157	1217	1285	1361	1445	1537	1637	1745	1861

Figure 1-19: Pythagorean triples created by Euclid's formula

The zigzagged partition line is provided around triples in which hypotenuse "**c**" does not exceed 1000

It seems that our thorough efforts - using Euclid's formula - to find two other Pythagorean Triples whose hypotenuse "**c**" is a factor of **325**, has accomplished its goal.

Two additional triples that are fit for the purpose have been extracted from the tabulation in figure 1-19 and are shown separately in figure 1-20.

m	n	"a" and "b"		c	Factor required to scale up "c" to 325	Factored a, b & c		
7	4	33	56	65	5	165	280	325
17	6	204	253	325	1	204	253	325

Figure 1-20

We establish now that the **seven** Pythagorean Triples shown in the tabulation of figure 1-21, have the least common hypotenuse length **c = 325** units

In algebraic notation:

$$36^2 + 323^2 = 80^2 + 315^2 = 91^2 + 312^2 = 125^2 + 300^2$$
$$= 195^2 + 260^2 = 165^2 + 280^2 = 204^2 + 253^2 = 325^2$$

s/n	a	b	c
Established by the Combined Process			
1	36	323	325
2	80	315	325
3	91	312	325
4	125	300	325
5	195	260	325
Added by Euclid's formula			
6	165	280	325
7	204	253	325

Figure 1-21: The seven Pythagorean Triples solving the Chapter Quiz

The solution of the Chapter Quiz is graphically shown in figure 1-22

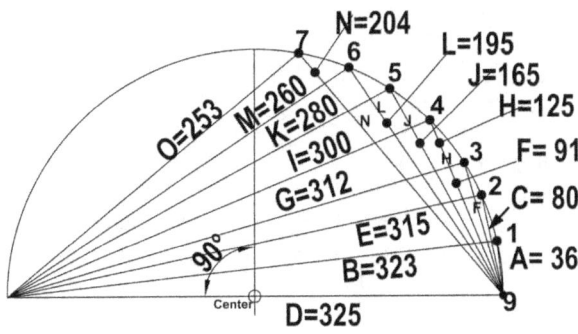

Figure 1-22: The solution of the chapter quiz

Chapter 2: The Network of Shortest Length

An experimental approach

Finding the shortest total length of line segments that connect a given set of points is an optimization problem that has boggled the minds of mathematicians and engineers.

The problem emerges in real life in several engineering applications such as designing networks of roads connecting several cities, routing of piping networks in a landscape or determining the shortest path of cables connecting several points.

Such situations - whereby a given set of points are interconnected by a set of line segments - is the concern of branch of mathematics called "Graph Theory". The diagram that represents a finite set of points interconnected by a set of line segments is called "Graph".

The Steiner tree problem – so called after Jacob Steiner, a Swiss mathematician who specialized in geometry – is concerned with the optimization of the total length of the connecting line segments, or the spanning tree in relevant terminology, that is finding the linking combinations corresponding to the shortest possible total length of line segments (the spanning tree).

To get the shortest lengths of roads connecting a given number of cities – for example - we may need to introduce additional – intermediate - points known as Steiner points.

For example, to minimize the total length of roads interconnecting three non-collinear cities (forming vertices of a triangle), we may have to add one Steiner point inside the triangle.

If we have four cities to be interconnected by roads of the minimum length then we generally add two more intermediate Steiner points.

Consider three points **A**, **B** and **C** forming the vertices of an equilateral triangle **ABC** having a side length "**S**". Let us examine the various combinations of connecting the three points.

In figure 2-1a, connection is directly made through **BA** and **AC** without introducing any Steiner points. Total connection length "L" equals **2 S**, while in figure 2-1b a single Steiner point "**D**" is introduced and connection is made through **AD** and **BC** with total connection length "L" equals (**S + 0.866 S = 1.866 S**) which is shorter from the case in figure 2-1a.

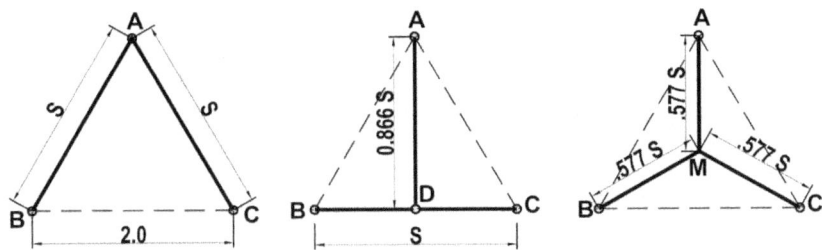

Figure 2-1a: L= 2 S   Figure 2-1b: L= 1.866 S   Figure 2-1c: L= 1.732 S

The connection in figure 2-1c is also made through one Steiner point taken in the triangle centroid, which is also its **incenter**, **circumcenter** and **orthocentre** for its being equilateral triangle. This latter connection – having a total length "L" of **3 x 0.57735 S = 1.73205 S** is apparently the shortest total connection length.

We just established from this foregoing discussion that the shortest path connecting three points **A**, **B** & **C** positioned at equal distances (as of the vertices of an equilateral triangle) is formed of three line segments (**AM**, **BM** & **CM**) intersecting at a fourth point "**M**" (known as Steiner point). Point **M** coincides with the centre of the triangle such that line segments **AM**, **BM** & **CM** are oriented at an equal angle of **360/ 3 = 120°** between them as shown in figure 2-1c.

Let us call these line segments (**AM**, **BM** & **CM**) forming the shortest path, "*the shortest path axes*".

This result is somehow expected, in view of the immaculate 3 axes symmetry of the equilateral triangle in the figure, but it will be thrilling to know that the shortest path axes **AM**, **BM** and **CM** will always be regularly positioned such that an angle of **120°** will separate one another, even in cases where the triangle **ABC** is irregular, provided always that none of its angles (**A**, **B** or **C**) equals or is greater than **120°**.

We shall come to this interesting conclusion later in our search for the shortest path.

Consider the three points forming the irregular triangle **ABC** in figure 2-2, and assume that line segments **AM**, **BM** and **CM** define the shortest path connecting points **A**, **B** and **C**.

Let us see what will happen if we slide down point **A** to the position marked **A'**.

Assume that the junction point where the shortest axes path meet (Steiner point **M**) will move to **M'** as a result of sliding **A** down to **A'**, so **A'M'**, **BM'** and **CM'** will constitute the *shortest path axes* for points **A'**, **B** and **C**.

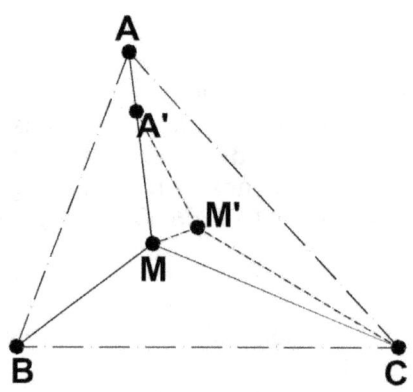

Figure 2-2

From this assumption we conclude that:

MA + MB + MC   < M'A + M'B + M'C,   (for MA, MB & MC being the shortest path axes connecting **A, B & C)**

M'A' + M'B + M'C < MA' + MB + MC,   (for M'A', M'B & M'C being the shortest path axes connecting **A', B & C)**

Now add the two respective sides of above two inequalities:

(MA + MB + MC) + (M'A' + M'B + M'C) < (M'A + M'B + M'C) + (MA' + MB + MC),

Now delete terms referring to line segments that are contained at both sides of the inequality:

MA + M'A'   < M'A + MA',

Replace MA by MA' + A'A

MA' + A'A + M'A' < M'A + MA',

Delete MA' from the two sides of the inequality

A'A + M'A' < M'A

But A'A, M'A' & M'A are three sides of a triangle, and since no single side of any triangle can have a length greater than the sum of the two other sides, the result indicated by this latter inequality is impossible. The conclusion is: assuming that junction point "M" will be shifted outside original shortest path axes (AM, BM & CM) as a result of sliding point A down to A' is an incorrect assumption.

Using the "Proof by Contradiction" technique we just reached to an interesting result that may be rephrased as follows:

In the regular triangle **ABC** of figure 2-1c, if you slide point **A** along **AM** down to another point **A'** (as we did in figure 2-2), the junction **M** (the Steiner point) of the shortest path axes will remain in its original position and the shortest path will be determined by line segments **A'M**, **BM** and **CM**. Upon sliding points **A**, **B** or **C** along the axes **AM**, **BM** or **CM** respectively, the triangle will not be regular any further and will take any proportions like that shown in the irregular triangle in figure 2-3

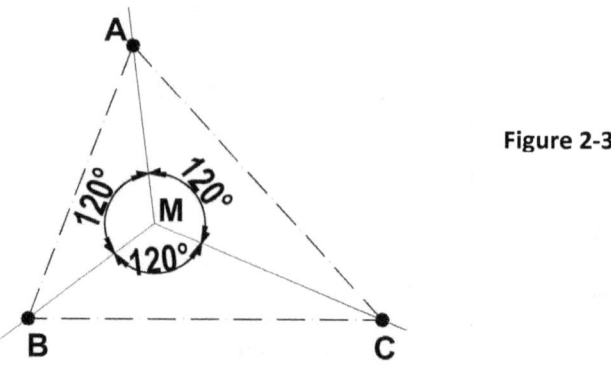

Figure 2-3

However, these same axes defining the shortest path in the case of regular triangle will still remain in position and an angle of **120°** will still be separating one another. It means that in the general case where an irregular triangle is considered, the junction "**M**" (the Steiner point) will be positioned such that the three shortest path axes will be having an angle of **120°** between one another as in the case in figure 2-3

The question now is how to construct the shortest path formed of the three lines connecting the given points **A,B** and **C** (see figure 2-4) where none of the angles at **A**, **B** or **C** equals or greater than 120°.

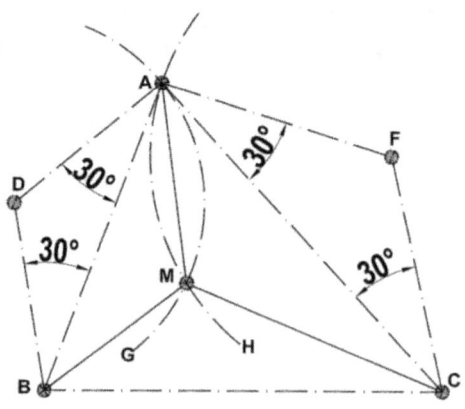

Figure 2--4

Here are the steps:

Draw the line segments **AF** and **CF** at an angle of **30°** with **AC**. Similarly, draw the line segments **AD** and **BD** at an angle of **30°** with **AB**. Having point "**F**" as a centre, draw the arc **AMH**. This arc is the locus of a point "**M**" such that peripheral angle **AMC** shall have the measure 0f **120°**, being half the measure of the central angle **AFC** (**240°**). Similarly, and having point "**D**" as a centre, draw the arc **AMG**. This arc is the locus of a point "**M**" such that peripheral angle **AMB** shall have the measure 0f **120°**, being half the measure of the central angle **ADB** (**240°**). The two arcs will meet at point "**M**". Now you have line segments **AM**, **BM** and **CM** constituting the shortest path connecting points **A**, **B** and **C**.

Apart from the analytical and logical study that has just established the rigidity of this **Y**-shaped Shortest Path Axes separated by **120°** between one another, I am presenting here an experimental approach that asserts the same result.

These experiments do not need any actual laboratory setup as it will be performed on paper. We shall be using - as a tool - the law of equilibrium of forces. Such on-paper-experiment is known as the "***Thought Experiments***". We shall be using it again in chapter 10 in which Einstein's Relativity will be presented. It is worth mentioning that Albert Einstein used it often to prove his Special and General theories of Relativity.

If we want to find the shortest path between two points, we can do that by extending a taut rope between them. Let us extend the same concept to cases of 3 and 4 fixed points.

Consider the three ropes **AM**, **BM** & **CM** connected at a junction M which is free to move by the ropes which are kept taut by three equal weights dropped from rollers into holes in the testing table as shown in figure 2-5a

An equal force will pull each of the free ends of the three ropes (**A**, **B** & **C**), and the junction point "**M**" will settle at an equilibrium position. As in the case of two points, we should also expect that ropes **AM**, **BM** and **CM** will have the shortest total length connecting points **A**, **B** & **C**.

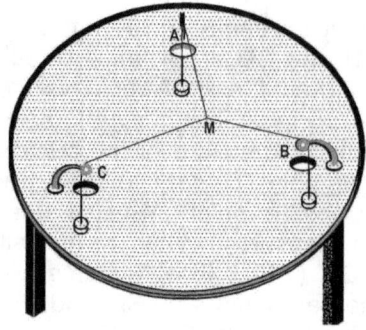

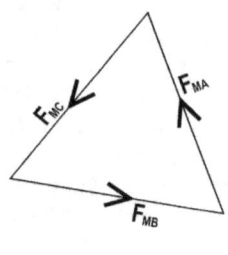

Figure 2-5a: Thought experiment to find the shortest path connecting three points

Figure 2-5b: The Force Triangle for the three forces acting at the junction "M'

Let us next find the angle between one rope and another in the state of equilibrium. To do so, we would draw the force polygon (here it is force triangle) for the puling forces in the three ropes F_{MA}, F_{MB} and F_{MC}. While these forces are equal in magnitude they obviously vary in direction. The force triangle (shown in figure 2-5b) shall have three equal sides – because of the equal magnitude of the forces it represents - but its sides will have to be parallel to the three ropes.

Since the three interior angles in any triangle will sum up to **180°**, each angle in such an equilateral triangle should measure **180°** /3 = **60°**. Exterior angle at each node should therefore be supplementary to **60°** (that is = **180°** - **60°** = **120°**). Since the three ropes in   figure 2-5a must be parallel to respective sides in the force triangle in figure 2-5b – as established earlier- we would expect that the ropes **AM**, **BM** and **CM** connected at point **M** will settle in equilibrium at an angle of **120°** separating one another.

We have just confirmed experimentally a conjecture that was deduced analytically asserting that the shortest path connecting three non-collinear points **A**, **B**, & **C** is formed of three lines meeting at a fourth internal point **M** (Steiner point) such that the angle between one line and another equals **120°**, provided always that none of the   interior angles of the triangle **ABC** equals or greater than **120°**

Let us now consider the shortest path connecting four points, starting from the special case where these points **A**, **B**, **C**, and **D**; are the vertices of a square.

In our example, the four points **A**, **B**, **C** and **D** form the vertices of a square having a side length of **2.0.**

In figure 2-6a, the four points are connected through three sides of the square **DA**, **AB** and **BC**, and the total connection length is therefore 3 x **2.0** = **6.0**.

In figure 2-6b the two diagonals of the square are linked with a single Steiner point "**M**" created. Total connection length thus equals: **2 x 2** $\sqrt{2}$ = **4 x 1.4142** = **5.657**, which is shorter than the connection in the arrangement of figure 2-6a.

The connection suggested in figure 2-6c adds a second Steiner point "**N**". While **M** is assumed to have an offset of "**X**" from the edge **AD** to the right side, **N** is given the same offset from the edge **BC** to the left side.

The connection length in this case will be:

$$L = 4 * \sqrt{1 + X^2} + (2 - 2X)$$

As in the case of connecting three points, we will find that the three angles at junction point **M** should have an equal magnitude of **360/3=120°**, in order for the total connection length to be the least possible (see figure 2-3). Obviously, the same is applicable to junction **N** in figure 2-6c. We are going to prove that now.

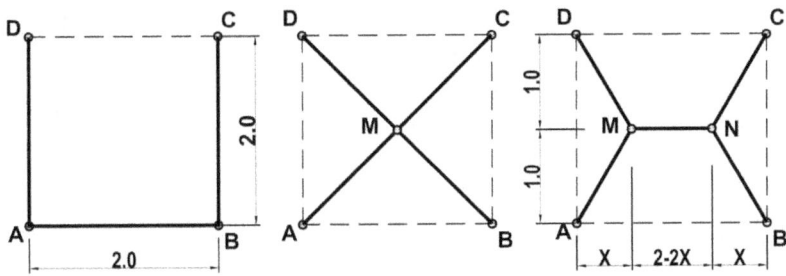

Figure 2-6a:

Figure 2-6b:
L = 5.657

Figure 2-6c: Finding the shortest "L"

33

If angle **AMD** in figure 2-6c is to be **120°** as discussed; angle **MAD** should be **30°**; in which case:

$$\frac{X}{1.0} = \frac{1}{\sqrt{3}} \quad \text{; hence } X = 0.57735 \quad \text{and}$$

$$L = 4.618802 + 0.845299 = 5.4641$$

This is even shorter than the connection length in the arrangement of figure 2-6b, and indeed the shortest possible connection length for points **A**, **B**, **C** and **D** as laid out in figure 2-6c.

Let us generalize our search for shortest path a little bit more by taking the four points **A**, **B**, **C** and **D** as vertices of a rectangle having the breadth of "**2 B**" and the height of "**2 H**" as that shown in figure 2-7a.

We should find out now whether angle **u** at the Steiner junctions **M** & **N** should be equal to **120°** if we are to establish the shortest connection joining the points **A**, **B**, **C** and **D**.

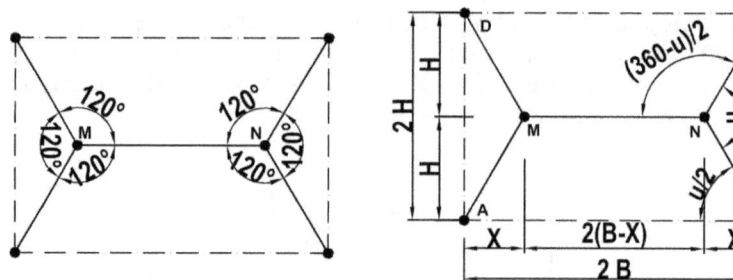

Figure 2-7a   **Figure 2-7b**

Total length of connecting lines:

$$L = 4\sqrt{X^2 + H^2} + 2(B - X)$$

$$L = 4H\sqrt{\left(\frac{X}{H}\right)^2 + 1} + 2H.\left(\frac{B}{H} - \frac{X}{H}\right)$$

The task now for us is to find the $\left(\dfrac{X}{H}\right)$ ratio that corresponds to the least length of "**L**" (the total connecting length). Let us remember that $\left(\dfrac{X}{H}\right)$ is nothing but the tangent of the angle **CBN**, in   figure 2-7a, and for convenience I shall replace $\left(\dfrac{X}{H}\right)$ in the above equation by **K** so we shall get:

In the above equation; **K** is an independent variable and **L** is a

$$L = 4H \sqrt{K^2 + 1} + 2H. \left(\frac{B}{H} - K\right)$$

dependant variable i.e. its value will depend on the value of **K**.

We know that calculus – and in particular differentiation - is the branch of mathematics conventionally concerned with evaluating

the rate of change of variables and finding their maxima and minima, so let us find the first derivative (the differentiation) of "L" in the equation above:

$$\frac{dL}{dK} = \frac{4H}{2\sqrt{K^2 + 1}} * 2K - 2H$$

$\dfrac{dL}{dK}$ Is the rate of change in total length "**L**" for an infinitesimal change in "**K**", and this rate converges to 0 where "**L**" is maximum

or minimum, so we need to equate $\dfrac{dL}{dK}$ to **0** to find the its minimum value

For $\dfrac{dL}{dK}$ =0, $2\,K = \sqrt{K^2 + 1}$, hence $3K^2 = 1$ or **K** = $\dfrac{1}{\sqrt{3}}$

The tangent of the angle **CBN**, (figure 2-7a) equals $\dfrac{1}{\sqrt{3}}$, therefore

∠**CBN = 30°**

Angle **ABN** (equals **U/2)** is complementary to angle **CBN** (the sum of the two angles = **90°**), hence ∠**ABN = U/2 = 90 - 30 = 60°**, hence **U= 120°**

As was the case with connecting three points and with connecting four vertices of a square, we find that: to connect the four vertices of a rectangle, the three lines at a Steiner point junction (points M, N in figure 2-7) should have an angle of 120° between one another in order for the total length of line segments connecting the 4 vertices would be the least possible

The linkage shown in figure 2-7a (**5** line segments) should be seen as typical and rigid in joining 4 rectangular vertices optimally, except that the length **MN** may be adjusted by contraction / extension and the four vertices may slide along – and never outside – the inclined lines, to suit the dimensions of any rectangle, as was the case in joining three vertices of the triangle in figure 2-2.

I know that some readers do not feel friendly with the differentiation analysis of the sort used in this example. For those readers, I have another option.

Let us try the experimental approach used in figure 2-5a for the optimum connection of three points. We will need the testing desk as that shown in figure 2-5a but with 4 holes positioned at the vertices of a rectangle as that shown in figure 2-8

You will notice that the two Steiner points in the middle of the testing table are represented by rollers in order for the tension forces to be consistent in the five segments of the rope and for the rope to be free to move and to naturally take the shortest path – as required by the experiment.

Forces in connecting ropes may be represented by the two force triangle shown in figure 2-9, for three forces meeting at the Steiner points **M** & **N** respectively.

Each of these points is in a state of equilibrium under the three tension forces acting on it.

Sides of the force triangle represent the forces in the three ropes that meet at a Steiner junction. These sides must be parallel to respective forces they represent (the three ropes) and their lengths ought to be proportional to the magnitude of these forces. Since the forces in the ropes are equal in magnitude, we should expect that the force triangle should be equilateral as shown in figure 2-9. And since any exterior angle between two adjacent members in an equilateral triangle equals **120°**, (being supplementary to the interior angle of **60°**) we should expect that the angle between each two ropes to also be **120°**,

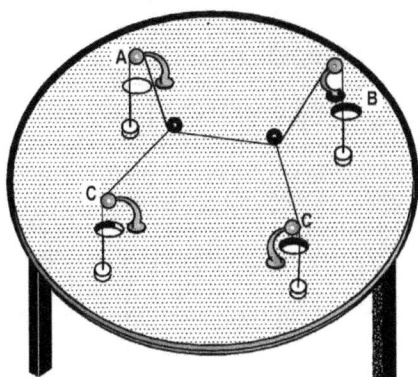

Figure 2-8: Thought experiment to find the shortest path connecting

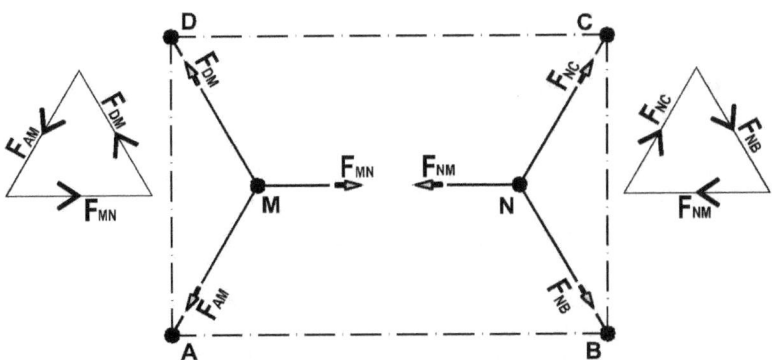

Figure 2-9: The Force Triangles representing the tensile force in ropes

From foregoing discussions, we conclude that the shortest path joining three points should pass through an additional Steiner point, and that the shortest path joining four points should pass through two additional Steiner points.

In all cases, Steiner points will be a junction at which three line segments will meet at an angle of **120°** between one another as typically illustrated – for the case of four points- in figure 2-10

Shortest path connecting points **A**, **B**, **C** & **D** in this case is marked by the line segments **AM**, **MD**, **MN**, **NB** & **NC**. We can slide point **A** along **MA'**, **B** along **NB'**, **C** along **NC'** or **D** along **MD'** inward or outward (figure 2-11) and still have the shortest path defined by the same line segments after relocating sliding points (similar to the case in figure 2-2 in which 3 points are connected).

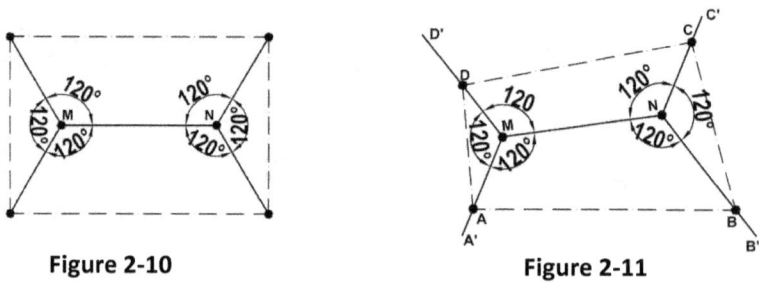

Figure 2-10   **Figure 2-11**

The only rigid constraint that should always be maintained is that the angle between one line segment and another in the path should remain **120°**.

Additionally, none of the interior angles **A, B, C or D** of the quadrilateral **ABCD** (or any polygon in case of having more than 4 points to connect) should be equal or greater than **120°**.

We should also know, that in some cases shortest path might not be unique, and you may have more than a path – of the same length- for the same setup of points as evident in the example in figures 2-12 and 2-13

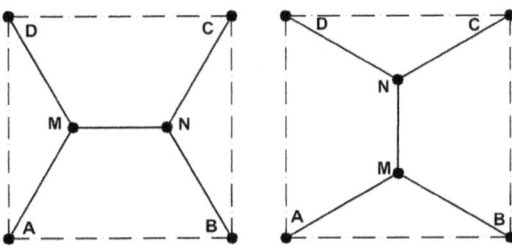

Figure 2-12   Figure 2-13

The question now is how to construct such a network having the shortest connection length to the four points **A**, **B**, **C** & **D** generally positioned as the vertices of an irregular quadrilateral.

Follow these construction steps illustrated in figure 2-14.

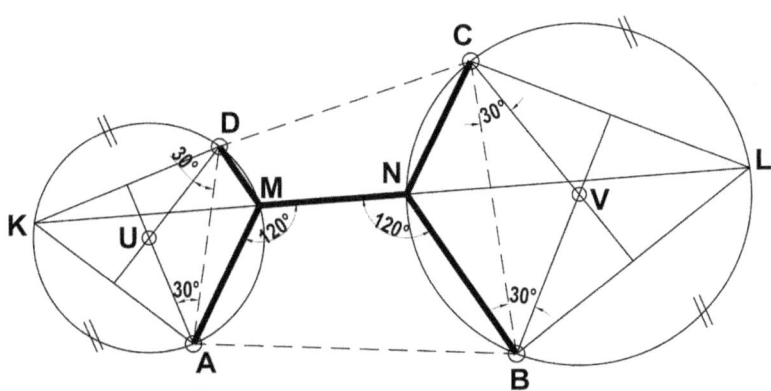

Figure 2-14

From the sides of quadrilateral **ABCD**; construct equilateral triangles on the two opposite sides of a shorter average length. In the figure, triangles are **BCL** and **DAK** are constructed on sides **BC** and **DA** being the shorter opposite sides (in average) of the quadrilateral. If you are not sure about which two opposite sides are shorter you will probably have to repeat the construction process on the two couples of opposite sides. From the centres **V** and **U** of the triangles draw circles **CBL** and **DAK**. Draw a straight line between **L** and **K**. It will meet circles **CBL** and **DAK** at points **N** and **M** respectively. **M** and **N** are the two Steiner points and the shortest path connecting points **A**, **B**, **C** and **D** comprises the bold black line segments **NB**, **NC**, **MA**, **MD** AND **MN**.

Before discussing the proof, I would like to remind the reader that if a central angle in a circle (**ADC** in figure 2-15) shares a common arc in it with a peripheral angle (**ABC**) then the measure of the central angle is as twice as the measure of the peripheral one.

Now back to the construction in figure 2.14. In circle **CBL**, the measure of the central angle BVC is **240°** because it equals **360°** - **120° = 240°**.

Any peripheral angle sharing the greater arc **CBL** but drawn in the smaller arc **BNC** will have the measure of **240/ 2 = 120°** and that is just the case with angle **BNC**.

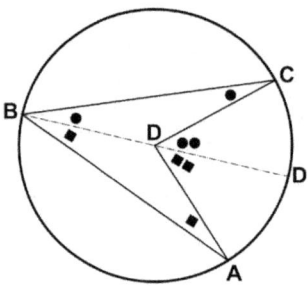

Figure 2-15

In view of the symmetry of the equilateral triangle about line segment **LV**, point **L** equally bisects the arc **BLC** into two equal arcs.

It follows that the peripheral angle CNL drawn on half the arc **CNB** will measure half the angle **CNB** i.e. ½ * **120° = 60°**. Angle **CNM** is supplementary to angle **CNL** (they sum up to **180°**), hence angle **CNM = 180 – 60 = 120°**. It follows that line segments **NC**, **NB** and **NM** are separated from one another by **120°**. In the same way we can also prove that angle **NMA** = angle **AMD** = angle **DMN** = **120°**.

I figure that with the technique and information gathered so far – especially from the experimental approach - we can move a step forward to the realm of five points. But let us only handle the uniform pattern in which the five points are vertices in a regular pentagon. Let us start with a logical deduction. When we were considering the shortest path connecting three points we had to introduce a single Steiner point. When four points were considered we had to add two Steiner points. We can confidently establish that three additional Steiner points would be required to define the network of the shortest length to connect five points. Based on the knowledge we have now, we would also establish that three line segments will be meeting at each Steiner point with an angle of **120°** separating each one from another.

We would also expect that each of the five points to be connected

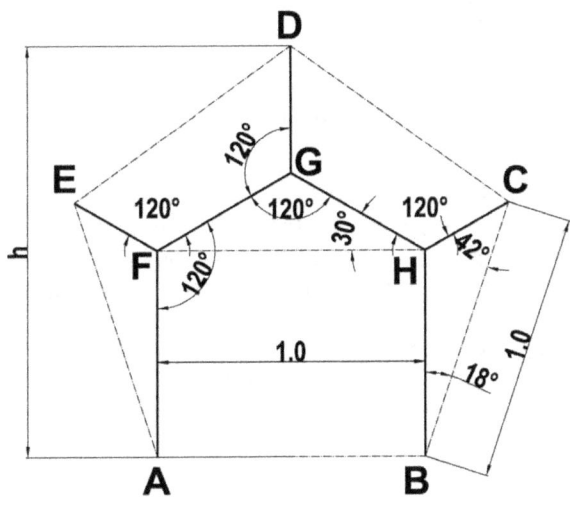

Figure 2-16

will be linked by a single line segment. In conclusion we would be having the connection network shown in figure 2-16.

Notice the three Steiner points introduced **F**, **G** & **H** and the Y-shaped connects positioned at these points.

Let us calculate the total connection length, assuming that length of each side in the regular polygon is **1.0**.

Applying the "sine law" on the triangle **HBC**

$$\frac{HC}{\sin 18} = \frac{HB}{\sin 42} = \frac{1}{\sin 120} = \left(\frac{2}{\sqrt{3}}\right)$$

From which HC = 0.1784, HB = 0.3863,

And applying the "sine law" on triangle **HGF**

$$\frac{HG}{\sin 30} = \frac{FH}{\sin 120}$$

Hence: $\dfrac{HG}{0.5} = \dfrac{1.0}{\sqrt{3}/2}$ from which **HG** = $\dfrac{1}{\sqrt{3}}$

DG = h – HB –HG. sin 30º where h is the polygon's altitude from the apex **D** to the midpoint of the base **AB** and HG. sin 30º is the altitude of the triangle **HGF**

$$h = \frac{1}{2\tan 18} = 1.53884, \; DG = 0.7195$$

Connection length = 2 (DG/2 + HG + HB + HC)= 3.581, which is less than what we will have if we connect four vertices of the polygon directly without introducing the three Steiner points **F**, **G**, **H**; at which case connection length would have been **4 x 1 = 4** instead.

Again, I must state that the shortest connection arrangement we got here is not unique. We can have four more similar arrangements having the same connection length by re-orienting the network as shown in figure 2-17

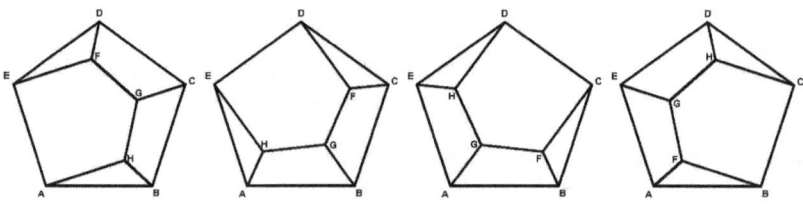

Figure 2-17

This will encourage us to move a step further to the regular hexagon. With the practice we just had in constructing the shortest networks we should find it easy to deal with the hexagon, and should expect it to just be as shown in figure 2-18

Intuitively, we may assume that there would be having four Steiner points, **G, I, H & J.** We can conclude that to join "**n**" points we will need to add **n-2** Steiner points.

There are 9 short line segments of equal length that make all the connections of the hexagon vertices in figure 2-18

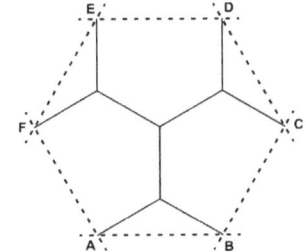 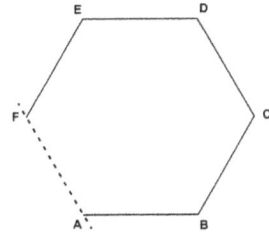

Figure2-18   **Figure 2-19**

Assuming the side length of the regular hexagon to be 1.0, the length of each line segment will be: $\dfrac{1}{\sqrt{3}}$ hence total connection

length will be $\dfrac{9}{\sqrt{3}} = 3\sqrt{3} = 5.19615$

It seems that we are in for a surprise!, because if we directly join the six points as shown in figure 2-19 without introducing the four Steiner points, we will – unexpectedly - be getting a shorter connection length of **5.0** only, so where is the catch?

I would repeat the warning raised earlier when the connection of four points was discussed that **none of the interior angles of the polygon joining the points to be connected should be equal or greater than 120°"**

The issue in the connection of figure 2-18 is that the interior angle in the regular hexagon - like the one we connect its vertices - is **120°**, which makes it unfit for the Steiner kind of connection we are trying now. This applies to all cases whereby more than five vertices -regularly arranged - are to be connected, because regular polygons having more than 5 sides will also be having interior angles measuring > **120°**

The Chapter Quiz

Triangle ABC (figure 2-20) is oblique, that is all its angles (A, B & C) are acute ($\angle$ 90°).

The three altitudes AD, BE and CF meet the sides of the triangle at D, E & F respectively. The triangle DEF is called the *Orthic triangle*.

Prove that the orthic triangle is the shortest connection between the three sides of triangle ABC following both conventional and the experimental approaches

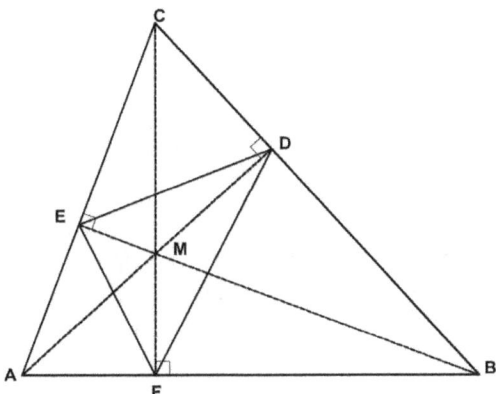

Figure 2-20

Discussion and solution of the Chapter Quiz

Though still concerned with the shortest path, this problem is different from the main theme discussed in this chapter in that the points to be connected are not identified but confined along given line segments. Instead of introducing additional points along the path (Steiner points), the question here is about identifying the position of the points along the given line segments.

The problem was first posed by **Fagnano** (1682-1766) (*Cajori 1909*).

Before getting into the proof details, I feel that two important concepts should be discussed as a starter.

First is the algorithm of identifying the shortest path between points **A** and **B** but after passing by a point on a given straight line as shown in figure 2-21

The second is to remind the reader with the properties of the Cyclic Quadrilateral i.e. the sort of quadrilaterals that can be circumscribed by a circle.

Consider this story: A cowboy wants to move from town "**A**" to town "**B**" but he should pass by the river line "**CD**" for his thirsty horse to drink from the river (figure 2-21).

Can you guide him to the shortest path?

Let **A'** be the mirror image of point "**A**" across the line **CD**, and assume that point **C'** is chosen for the horse to drink. The path will be along **AC'** and **C'B**. Since **AC'** = **A'C'** (for **A'** being the mirror image of **A**) we can establish that the length of the trip in this case will be the length of the broken line **A'C'** - **C'B**. Since it is a broken line, it cannot possibly be the shortest path, so we should join **A'B** through a straight line **A'B** that meets **CD** at **C''** to get the shortest path.

Shortest path is therefore defined by line segments **AC"** and **C"B**. This is the path of light beam emitted from a light source at **A**, reflected on a mirror **CD** and received by a vision detector **B**.

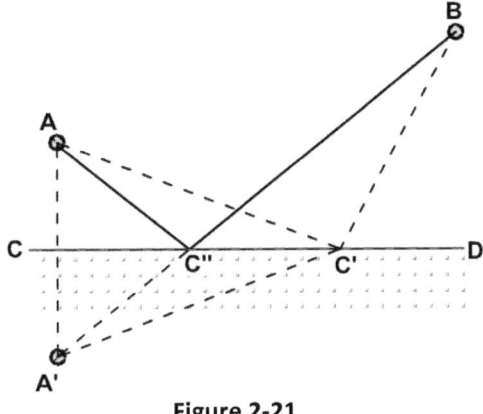

Figure 2-21

Second primer is a reminder of the properties of the Cyclic Quadrilateral

While any triangle can be circumscribed by a unique circle, this is not the case with all quadrilaterals. Only quadrilaterals of a certain configuration can be circumscribed by circles.

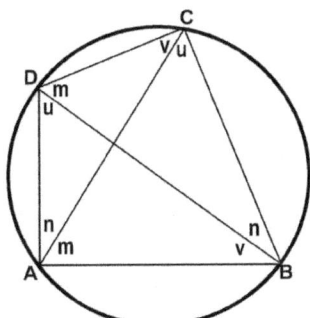

Figure 2-22

The condition that qualifies circumscription is simple: In order for a quadrilateral to be circumscribed by a circle, each two opposite angles in the quadrilateral should sum up to **180°**, in which case the quadrilateral will be called: "**Cyclic Quadrilateral**".

In figure 2-22, quadrilateral **ABCD** is circumscribed by a circle. There is no doubt therefore that **ABCD** is a cyclic quadrilateral, and the results of that will be:

- Each two opposite interior angles in the quadrilateral sum up to **180°**. In the quadrilateral in figure 2-22:

∠**ADC** + ∠**ABC**= 180, ∠**DAB** + ∠**DCB** = 180
Furthermore, if any of the quadrilateral's interior angles – say **DCB** - is a right angle (**90°**); the opposite interior angle (**DAB**) should also be right angle and the opposite diagonal (**DB**) will be the diameter of circumscribing circle.

- Peripheral angles sharing the same arc of a circle are congruent. This means that ∠**DAC** = ∠**DBC**, ∠ **CDB** =∠ **CAB**, ∠**BDA** = ∠**BCA** and ∠**ABD** = ∠**ACD**

Let us now proceed with the solution of the quiz.

We shall follow the "***proof by contradiction***" concept.
Assume that point **F** is positioned along the triangle's side **AB** randomly as shown in figure 2-23, where **CF** is not an altitude of triangle **ABC**

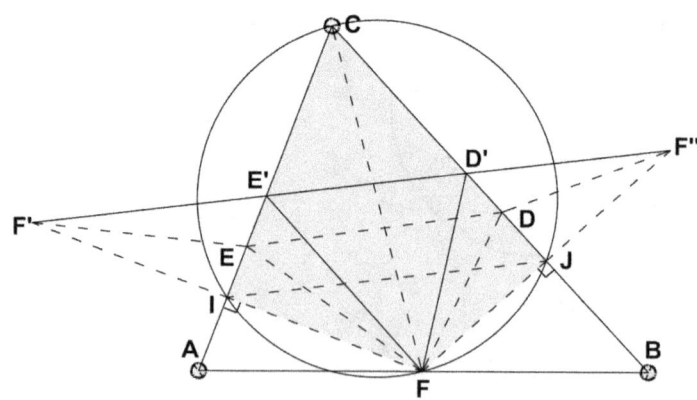

Figure 2-23

48

F' and **F''** are the mirror images of **F** at lines **AC** and **BC** respectively. If **E** and **D** are the other two vertices of the shortest path triangle (to avoid calling it **Orthic** till it is proven), the length of the path will be: **ED + DF + FE**.

We know that **DF = DF''** and **FE = F'E for F''** and **F'** being mirror images of **F** at **BC** and **CA** respectively.

Length of the shortest path therefore equals the length of the broken line **F''DEF'**. Since we know that the straight line (not the broken one) is the shortest length between two points, we should realize that the straight line **F'E'D'F''** is the one that should represent the shortest path connecting points on the sides of triangle **ABC**, if **F** is one of these points. The other two points in this case should be **D'** and **E'** and the shortest path will be the perimeter of triangle **E'FD'**.

The challenge will be on us to dispute the assumption that point **F** is correctly guessed, so let us wrap up the information at hand.

According to foregoing assumption, the shortest path – which is the perimeter of triangle **E'FD'**, also equals the length of line segment **F'F''**. **F'F''** is double the measure of line segment **IJ**; because **IJ** bisects two sides of triangle **FF'F''** in which **F'F''** is the base, so let us find the configuration that makes the length of **IJ** the shortest possible.

Remember that it is all about the correctness of our assumption that point **F** is a vertex in the shortest path triangle.

Consider the shaded quadrilateral **CIFJ** (figure 2-23). This is a cyclic quadrilateral because two of its opposite interior angles (**FJC & CIF**) are right angles and **CF** is a diameter of the circle that circumscribes it (as discussed in the introductory notes).

Furthermore **IJ** is a chord in this circle as seen in figure 2-23. Remember that the measure of **IJ** equals half the perimeter of the presumably shortest perimeter triangle joining the sides of triangle **ABC**. The chord **IJ** corresponds to angle **ACB**; which is a fixed angle. Let us follow a logical thinking sequence:

- The question is whether the selection of the position of point **F** would result in bringing the perimeter of triangle **E'FD'** to its absolute minimum.

It may be rephrased to read: Can we move point **F** along line segment **AB** such that **IJ** (a chord in circle **CIF** whose length equals half the perimeter of triangle **E'FD'**) will have the least possible length

- To shorten the chord IJ; we have to either reduce its peripheral angle ACB, which is not possible because it is fixed (being an interior angle in the triangle ABC), OR to reduce the size of the circle - namely its diameter.

  - The only way for the diameter CF of the circle to be reduced is to move point F such that CF becomes perpendicular to AB (i.e. CF becomes an altitude)

Similarly and by applying the same procedure on the other two sides of triangle **ABC**; we will reach to the same conclusion that the **Orthic** triangle (joining the foot points of the altitudes) is the triangle whose perimeter equals the shortest path between the three sides of rectangle **ABC**

Having just solved the Chapter Quiz using a traditional algorithm, let us address the problem using an Experimental Approach similar to that employed earlier in the cases of connecting three and four points. However, we have to apply a few changes in our lab setting. We have to allow for points **D**, **E** and **F** to move freely along line segments **BC**, **CA** and **AB** respectively (figure 2-24). How can we do that? This is a simple task.

We shall simulate the rigid lines **AB**, **BC** and **CA** by curtain rails, and simulate the freely moving points **D**, **E** and **F** by roller wheels to slide freely along the rails (and never outside it). In order for the three connection lines between the roller wheels to take the shorter length; they may be simulated by a single closed strongly taut rubber band to engulf the three rollers and push them to the position of minimum length for the band. The rubber band will take the shape of triangle **DEF** (figure 2-24)

When the rubber band is tightened about the three roller wheels they move along the curtain rails and settle at a position corresponding to the shorter length of the rubber band.

Assuming that the tensile force in the rubber band is **T**, the forces acting at each roller wheel are shown at the right hand side of the figure

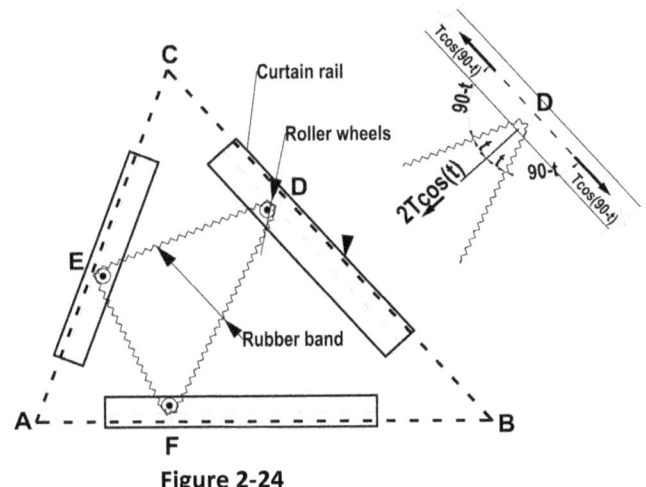

Figure 2-24

The rubber band should make an angle with the curtain rail before it turns around the roller wheel as it does after turning. The measure of this angle at point **D** is (**90 - t**) as shown in figure 2-24, where t is the angle between the band and the perpendicular to the rail.

That is because the roller wheel is balanced in the direction of movement along the rail under two components of forces T (before and after turning around the roller wheel) along the rail. These components are of the same magnitude: **T * sin(t)** but act in the opposite directions along the rail which causes the force in the direction of the rail to vanish, otherwise the roller wheel would keep moving

The sum of the two components of the tension force **T** in a direction perpendicular to the rail equals

2T cos(t) = 2T sin(90 - t).

This is the resultant force the rubber band will be exerting perpendicularly on each of the three curtain rails at the position of the roller wheels, where **t** is the angle between the rubber band and the line perpendicular to the rail.

The three forces acting on the triangular frame by the rubber band and the force triangle that represents them are shown in figure 2-25.

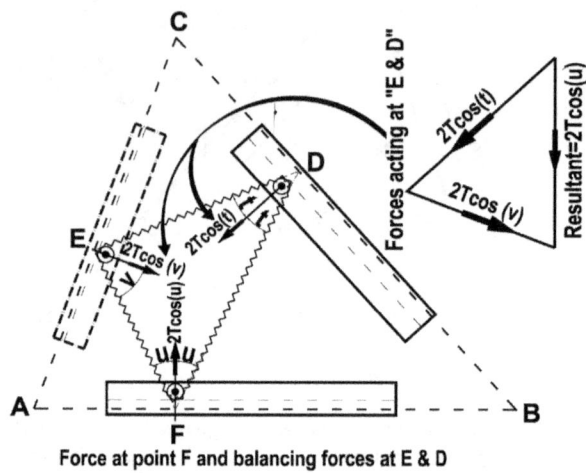

Force at point F and balancing forces at E & D

Figure 2-25

The two forces **2T*cos(t)** and **2T*cos(v)** acting at points **D** and **E** respectively (shown in figure 2-25) can be dissolved in the two directions along **CA** and **CB** in line with the force triangles chart shown in figure 2-26.

I shall be using the sine rule to get the magnitude of these two resolved forces.

Figure 2-26 shows the force triangle for forces exerted by the rubber band on the curtain rails and the dissolved equivalent forces acting along the curtain rails **AC** and **CB**.
The latter set of two force components is just an equivalent transformation to the former set.

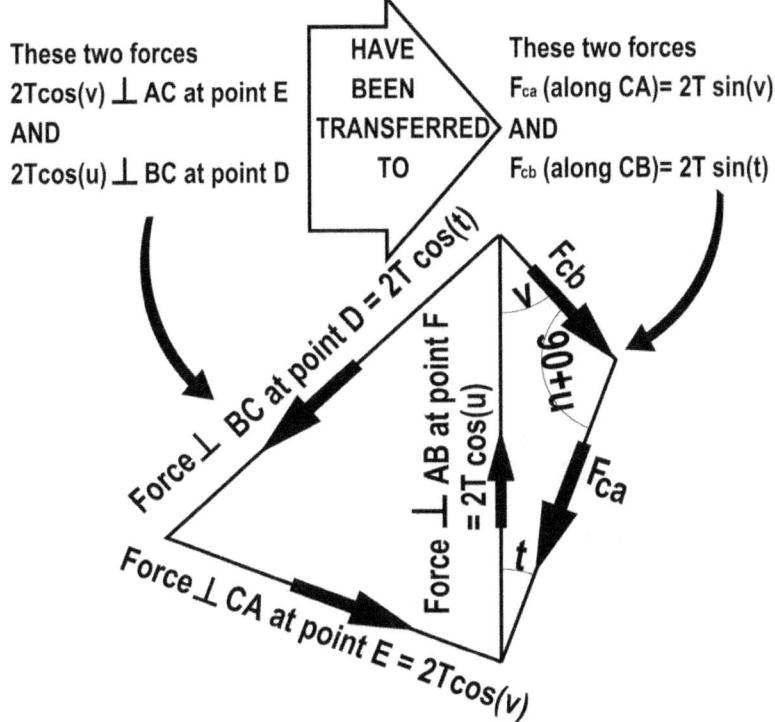

These two forces

2Tcos(v) ⊥ AC at point E

AND

2Tcos(u) ⊥ BC at point D

HAVE BEEN TRANSFERRED TO

These two forces

Fca (along CA)= 2T sin(v)

AND

Fcb (along CB)= 2T sin(t)

Figure 2-26

Applying the sine rule on the right side force triangle:

$$\left(\frac{Fcb}{\sin(t)}\right) = \left(\frac{Fca}{\sin(v)}\right) = \frac{2T\cos(u)}{\sin(90+u)}$$

Since sin(90+u) = cos(u), hence $\left(\frac{Fcb}{\sin(t)}\right) = \left(\frac{Fca}{\sin(v)}\right) = \ 2T$

from which we deduce that:

Fca = 2T sin(v) and Fcb = 2T sin(t)

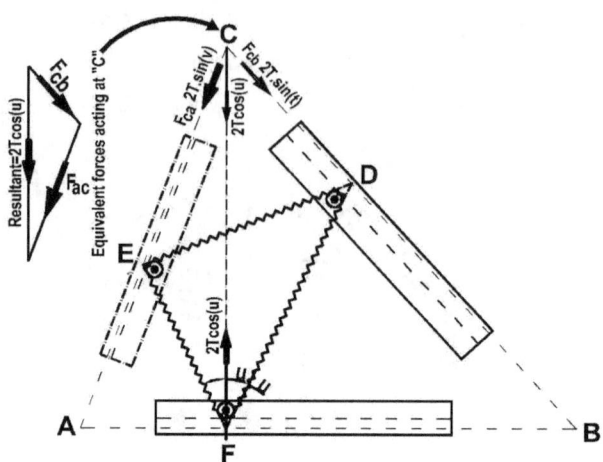

Force at point F and equivalent balancing forces along rails CA & CB

Figure 2-27

Figure 2-27 shows the balanced forces acting on points **C** and **F**. The force exerted by the rubber band will locate the three wheel rollers **D, E & F** at positions such that the band will have the least connection length **DE + EF + FD**.

Since the triangular frame is in equilibrium under the force **2T.cos(u)** (acting perpendicularly at point **F** of the side **AB**) and the two forces F_{ca} and F_{cb} acting along **CA** and **CB** respectively, the three forces must meet at a single point, and this point should be "**C**" being the meeting point of F_{ca} and F_{cb}. This means that point **F** must be the foot of the altitude **CF** on the triangle base **AB**.

Similar analysis can be applied on points **D** and **E** to assert that the Orthic triangle **DEF** is the shortest connection between the three sides of triangle **ABC**

However, we still have to prove that the three altitudes of triangle **ABC** are the bisectors of the interior angles of the shortest path triangle **DEF**.

This was established earlier as a result of the rubber band experiment; as explained in figure 2-24 and associated discussions, but yet to be proven geometrically.

Recall the discussions related cyclic quadrilateral (figure 2-22), and about peripheral angles sharing an arc of a circle.

Quadrilateral **FBDM** in figure 2-28 is cyclic i.e. can be circumscribed by a circle because each two opposite angles in it sum up to **180°** ($\angle$**MDB** = $\angle$**MFB** = **90°**). Angles **FBM** and **FDM** are peripheral and share the arc **MF** in the circle **FBD**, hence they are congruent: $\angle$**FBM**= $\angle$**FDM** = **t**.

Similarly, quadrilateral **EMDC** is cyclic because $\angle$ **MDC** = $\angle$**MEC**=**90°** hence $\angle$**MDC** + $\angle$**MEC** = **180°**. Angles **ECM** and **EDM** are peripheral and share the arc **EM** in the circle **CEM**, hence they are congruent ($\angle$**EMC** = $\angle$**EDM**) . Triangles **CAF** and **BAE** are similar (have the same proportions and interior angles) because they are right angled and have angle **CAB** common, therefore angle **FCA** = angle **ABE** = **t**. From foregoing discussion, $\angle$**FDA** = $\angle$**EDA** .

In the same manner, we can prove that $\angle$**EFC** = $\angle$**CFD** and $\angle$**FEB** = $\angle$**BED**

Here is the conclusion:

The shortest path of line segments connecting points on the sides of a triangle **ABC** is the **orthic** triangle **DEF**.

Point **M**, the **orthocentre** of triangle **ABC** (at which the three altitudes meet); is also the **incenter** (the intersection point of the bisectors of interior angles) of its **orthic** triangle **DEF**

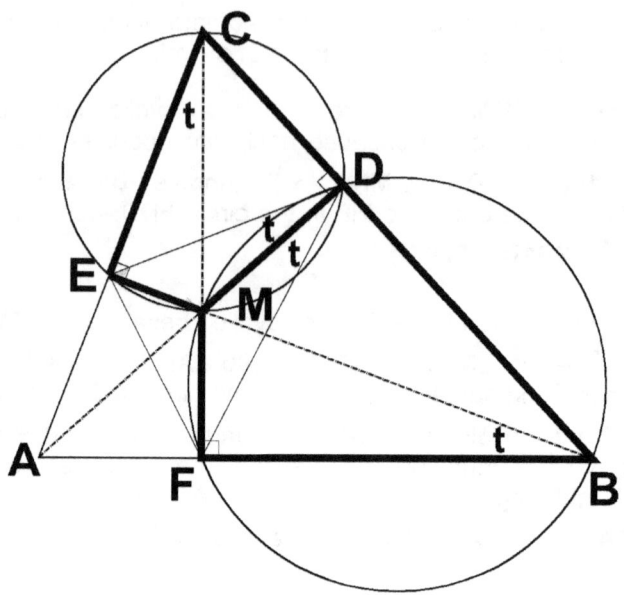

Figure 2-28

Chapter 3: The Golden Ratio and Fibonacci sequence

In his encyclopedic reference book "*The Elements - Proposition VI–30*", **Euclid** (365-300 BC) referred to dividing a line segment into a longer and shorter parts such that the longer to the shorter ratio is the same as the ratio between the whole segment length to the larger part.

Such a witty division of the line segment was described by Euclid as dividing the line segment into **extreme** and **mean** ratio.

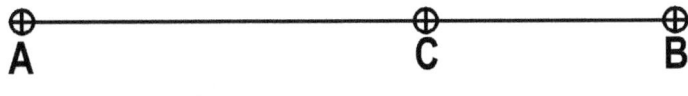

Figure 3-1

Point "**C**" is inserted in the line segment **AB** (figure 3-1) such that:

$\dfrac{AC}{BC} = \dfrac{AB}{AC}$ hence the line **AB** is said to be divided into extreme

and mean ratio.

In the 20th century, this ratio has been given other several common names such as "**Golden Ratio**", "**Golden Section**" and "**Divine Ratio**".

Let us find this ratio numerically. Assuming that the ratio is

$x = \dfrac{AC}{BC} = \dfrac{AB}{AC}$, Since **AB = AC + BC**

$\dfrac{AC}{BC} = \dfrac{AC+BC}{AC} = 1 + 1/(\dfrac{AC}{BC})$

hence $x = 1 + \dfrac{1}{x}$,

Multiplying the terms of the equation at the two sides by "**x**", we get:

$x^2 = x + 1$, or $x^2 - x - 1 = 0$,

This is a quadratic equation that can be solved using the general formula: $x = \dfrac{-b \pm \sqrt{b^2 - 4ac}}{2a}$ where **a = 1**, **b = - 1** and **c = -1.**

Substituting these values in the above formula we get the following two roots for x,

$$x = \frac{1 + \sqrt{1+4}}{2} = \frac{\sqrt{5}+1}{2} = 1.6180339 \ldots.$$

and $\quad x = \dfrac{1 - \sqrt{1+4}}{2} = \dfrac{-\sqrt{5}+1}{2} = - 0.6180339$

The **Golden Ratio** is however known to be a positive irrational number hence the negative root should be ignored. The Golden Ratio was given the Greek letter φ (pronounced phi) at lower case as a name at the beginning of the 20th century, so we may replace **x** by φ in above basic formulae such that it reads as follows:

$$\varphi^2 = \varphi + 1 \quad \text{where the Golden Ratio } \varphi = 1.6180339$$

A rectangle proportioned by length to breadth ratio of $\varphi = 1.6180..$ is called **Golden Rectangle**.

If a square having the side length equals the shorter side of a Golden Rectangle is drawn on that side of the rectangle; the remaining part of the rectangle will also be golden, that is will have the length : breadth ratio of **1.6180.. : 1.**

We can split it further to a smaller square and a smaller golden rectangle. The process of splitting a golden rectangle into a square and a smaller golden rectangle can be repeated several times till the square and rectangle become invisibly small as shown in figure 3-2.

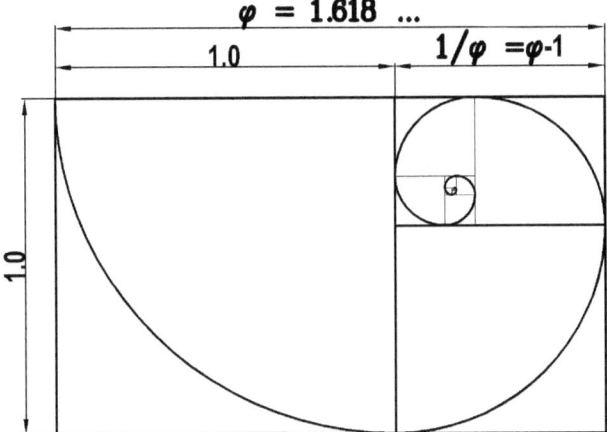

Figure 3-2

Notice the **Golden Spiral** which is formed by drawing a quarter of circles inside the squares created in the splitting process.

To construct a golden rectangle, start by drawing a square **ABCD** with side length equaling the required breadth of the golden rectangle (will assume it here to be 2.0) as shown in figure 3-3.

Second step is to bisect the square side **AB** at **E**.
With **E** taken as a center; draw an arc **CF** having **EC** as a radius and point **F** will be at the extension of **AB**. Complete the rectangle **AFGD**.

This is the required golden rectangle since its length will be equal

to $\sqrt{5} + 1$, hence length/ breadth ratio will $\dfrac{\sqrt{5}+1}{2} = \varphi$ be

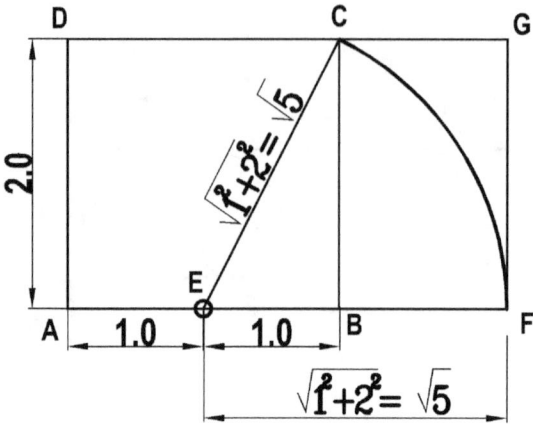

Figure 3-3

The great German mathematician and astronomer Johannes Kepler (1571 – 1630) said once that there were two treasures in Geometry:

the **Pythagorean Theorem** and the **Golden Ratio**, and while the former could be compared to gold; the latter might be compared to a precious jewel *(Stakhov 2003).*

I agree unreservedly with Kepler; but would like to add a third jewel: **pi**, the circle's circumference-to-diameter ratio (often written in Greek letter π).

Some aspects of the Pythagorean Theorem – related to Number Theory- were discussed in Chapter One. Pi will be the subject of Chapter Six and some of the treasures of the Golden Ratio φ are exhibited here.

I shall start with the geometric features of φ.

The proportions of the Golden rectangle (figure 3-2) have attracted many geometers, architects and artists for the inherent natural visual and aesthetic beauty in it.

The Golden Triangle is no less attractive.

It is an isosceles triangle in which the side to base ratio is the Golden Ratio φ. The Golden triangle is shown in figure 3-4. If the base angle in a Golden Triangle is bisected; a smaller Golden Triangle is created (linear ratio between the two triangles is φ).

You may continue creating smaller Golden Triangles in the same manner endlessly or until the size of emerged triangles becomes invisibly small.

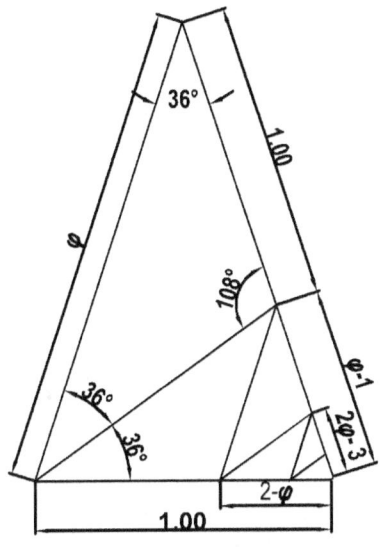

Figure 3-4

It is interesting that the relationship between the Golden Ratio φ and π (the circle's perimeter-to-diameter ratio) is embodied in the Golden Triangle and can be extracted from its geometry as follows:

The ratio of the base of the triangle to its side length

$= 1/\varphi = (\varphi -1) = 2 \sin 18° = 2 \sin (\pi /10)$, hence

$(\varphi -1)/ 2 = \sin (\pi /10)$ or $\qquad \pi = 10 \sin^{-1} (\frac{\varphi -1}{2})$ and may

also be expressed as $\qquad \pi = 10 \arcsine(\frac{\varphi -1}{2})$

The Golden Triangle is a basic component of the pentagram; which is also a deeply golden figure - if I may say.

Figure 3-5 shows a pentagram inscribed in a regular pentagon.

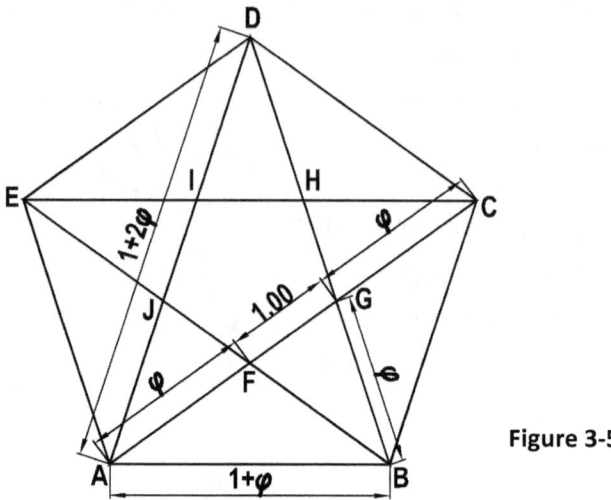

Figure 3-5

The presence of the Golden Ratio φ in the pentagram or in the pentagon is amazingly strong. Figure 3-4 is self-explanatory in exhibiting this fact; and it is sufficient to note that:

$$\frac{DH}{HI} = \frac{DG}{DH} = \frac{DC}{DH} = \frac{DB}{AB} = \varphi$$

Amongst regular polygons; or 2D objects in general; the regular pentagon and the pentagram are indeed most associated with the Golden Ratio.

In the realm of 3D geometry; the Golden Ratio emerges surprisingly from where one does not expect.

The regular polyhedra are three dimensional solids made up of regular polygonal faces. \a regular polyhedron is analogous to the regular polygon in 2D geometry.

There exist only five regular polyhedra (also called Platonic Solids) which will be discussed in chapter 4, but I shall only focus here on the appearance of the **Golden Ratio** in association with two of them: the *Icosahedron* and the *Dodecahedron*

The icosahedron is formed of 20 faces of equilateral triangles; with each five faces meeting at one of the icosahedron twelve vertices. The icosahedron can be constructed on a skeleton of three Golden Rectangles perpendicularly positioned one on another in the three orthogonal directions as shown in figure   3-6. The Golden Rectangles in the figure are:

ABCD (in plane **XY**), **EFGH** (in plane **YZ**) and   **IJKL** (in plane **ZX**)

The twelve vertices of the three golden rectangles are the vertices of the icosahedron. At each of these vertices; five of its equilateral triangular faces will be meeting. The five triangular faces that meet at vertex **H** are shown in the figure in dashed lines. The measure of the side of these equilateral triangles is equal to the shorter length (the breadth) of the golden rectangles. In the figure this breadth is assumed to be 2a. As you have seen, we construct the icosahedron by joining the twelve vertices of the three golden rectangles

The process is reversed in the case of the dodecahedron.

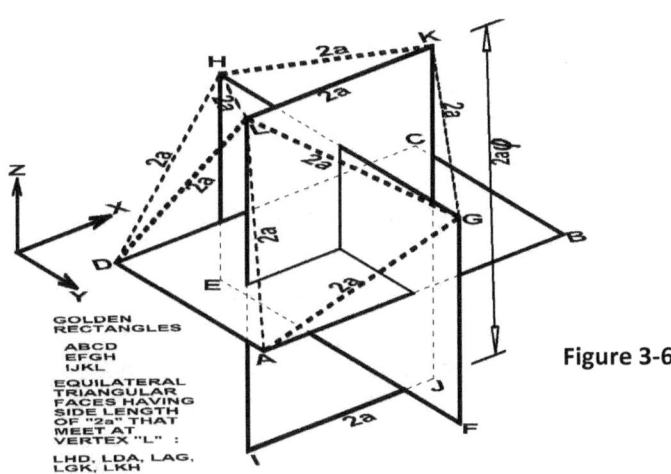

Figure 3-6

The dodecahedron is formed of twelve faces. Each face is a regular pentagon. Apart from the strong association between the regular pentagon and the golden ratio – as demonstrated in figure 3-5; there is an additional relation between the dodecahedron and the golden rectangles.

The centers of the twelve pentagons (the faces of the dodecahedron) are in fact the vertices of three golden rectangles perpendicularly positioned one on another in three orthogonal directions. So, instead of joining the vertices of the golden rectangles to construct the faces of the polyhedron – as was the case of the icosahedron, we join the centers of the regular pentagonal faces of the dodecahedron in three groups of four, to get three golden rectangles orthogonally positioned one to the other. This is demonstrated in figure 3-7.

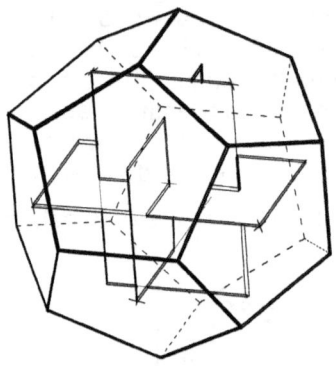

Figure 3-7

Another situation, in which the golden ratio emerges unexpectedly, is illustrated in figure 3-8.

If the sphere and open ended (bottomless) cone shown in figure

3-8 are to have the same height "**h**" and the same surface area, then the radius of the cone base and its slanted length should have the following measures respectively: $h/\varphi^{1/2}$ and $h*\varphi^{1/2}$.

This can be proved by applying the Pythagorean Theorem to confirm the radius of the cone base ($R_{cone} = h/\varphi^{1/2}$) then by substituting the values of the sphere diameter "**h**", the cone base radius R_{cone} and the cone slanted length L_{cone} in the following surface area formulae:

Surface area of the sphere $= \pi . h^2$

Surface area of the cone $= \pi . R_{cone} . L_{cone}$

You may notice that in the case in question, $R_{cone} \cdot L_{cone} = h^2$; hence the surface area of the sphere equals that of the cone.

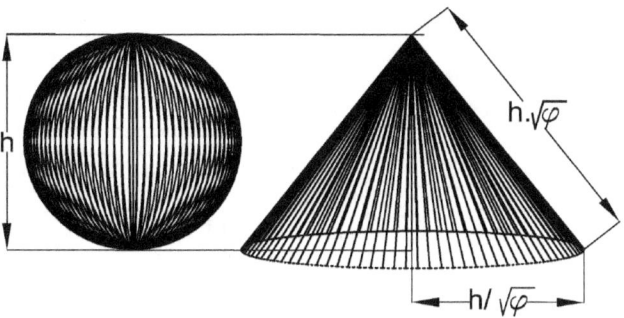

Figure 3-8

The special algebraic features of the Golden Ratio φ stem from the basic equation discussed earlier: $\varphi^2 = \varphi + 1$, because it simply means that the Golden Ratio φ is the number that you can square it just by adding 1 to it:

$$\varphi^2 = 1.6180..^2 \text{ and } = 2.6180..$$

We can continue further to find φ raised to greater powers. An interesting feature in φ raised to any power is that it can be reduced as in the above equation and expressed in terms of φ^1.

For example:

$\varphi^3 = \varphi * \varphi^2 = \varphi (1+ \varphi) = \varphi + \varphi^2$ and this latter expression equals $2\varphi + 1$

Above equation can be further generalized as follows:

$$\varphi^n = \varphi^{n-1} + \varphi^{n-2},$$

This interesting equation states that φ raised to a power **n** equals φ raised to power (**n-1**) added to φ raised to power (**n-2**), and here is the proof:

$$\varphi^n = \varphi^{n-2} \cdot \varphi^2 = \varphi^{n-2} (\varphi +1) = \varphi^{n-1} + \varphi^{n-2}$$

The basic equation $\varphi^2 = \varphi + 1$ may be written as $\varphi = \sqrt{1 + \varphi}$ and that φ under the root of the latter expression can be substituted by $\sqrt{1 + \varphi}$ and so on. With several repetition of the process we will get the following expression:

$$\varphi = \sqrt{1 + \sqrt{1 + \sqrt{1 + \sqrt{1 + \sqrt{1 + \sqrt{1 + \cdots}}}}}}$$

Another important result would also stem from the basic equation $\varphi^2 = \varphi + 1$, and by dividing the three terms by φ we get: $\varphi = 1 + 1/\varphi$ or $(1/\varphi) = \varphi - 1$

Again we have an interesting formula that simply says: "to get the reciprocal of φ, simply subtract 1 from it:

$$\varphi = 1.6180.. \quad \text{and} \quad (1/\varphi) = 0.6180..$$

In the equation $\varphi = 1 + 1/\varphi$ if we keep substituting that φ at the right side of the equation by $\varphi = 1 + 1/\varphi$ we will get the following repetitive pattern:

$$\varphi = 1 + \cfrac{1}{1 + \cfrac{1}{1 + \cfrac{1}{1 + \cfrac{1}{1 + \cfrac{1}{1 + \cdots}}}}}$$

The balanced appearance of the Golden Ratio have been appreciated and felt pleasing to the commons and have inspired intellectuals of artists and architects of all ages. Researchers sensed a strong association for the Golden Ratio with great architectural and artistic works.

Most renowned of these works are the great pyramid of Khufu in Egypt, the Parthenon, the great Greek temple of Athens, the renowned Mona Liza of Leonardo Davinci and many others.

Stakhov (2003) says that the great pyramid of Khufu in Egypt is one of the most outstanding works that exhibit the proportions of the Golden Section. **Burton (2011)** explains further by telling the story that has crept into recent literature. It says that Egyptian priests told Herodotus (a prominent historian of ancient time) that the dimensions of the pyramid were chosen such that the area of each face would be the same as the area of a square having sides equal to the pyramid's height. Figure 3-9a shows the pyramid height "**v**", its base side length "**2a**" and slanted face height "**h**". The dashed triangle in figure 3-9b is right angled that has horizontal side "**a**" vertical height "**v**" and hypotenuse "**h**", hence:

$h^2 = a^2 + v^2$ (Pythagorean Theorem)

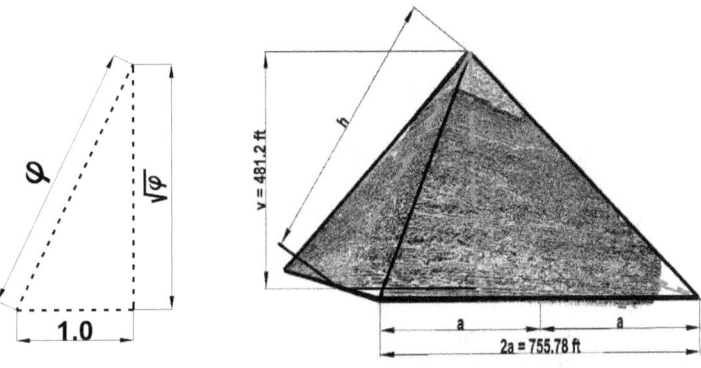

Figure 3-9b

Figure 3-9a

Area of each face = ah and the area of the square having its sides equal to the pyramid's height is v^2. According to Herodotus story:

$v^2 = ah$

Substitute v^2 by ah in the equation ($h^2 = a^2 + v^2$)

$h^2 = a^2 + ah,$

Now divide each term in the equation by a^2:

$$\left(\frac{h}{a}\right)^2 = 1 + \frac{h}{a}$$

Above equation says that in order for you to square the ratio $\left(\frac{h}{a}\right)$, simply add 1 to it.

This is pretty identical to the definition of the golden ratio φ and above equation is a replica of the following "φ" basic equation discussed earlier:

$$\varphi^{2} = 1 + \varphi$$

In conclusion, and according to foregoing calculations, the ratio

(slanted altitude of the pyramid's face) / (half the base side length):

$$= h / a = \varphi$$

Burton (2011) also asserts that the pleasing proportions of the

Golden Ratio has been widely implemented in the façade of the

Parthenon, that great temple of Athens.

Stakhov (2003) says that: "*The main cause of Parthenon's beauty is the exclusive harmony of its parts based on the golden proportion*"

Among several Golden and harmonic aspects that privilege the design of the great monument; the proportions of the main façade is shown in figure 3-10.

Figure 3-10

The Golden Ratio concept and pleasing perception are not limited to human creation. Several researchers claim that many natural features follow the golden proportions and the Fibonacci numbers. What are the Fibonacci numbers?. We are coming to that soon.

The human body is said to be exhibiting the Golden Ratio in several aspects. While I recognize these natural associations with the divine number, I would focus on the mathematical aspects of the Golden Ratio, rather than debating on philosophical theories supporting a concept viewing that nature –also- appreciates the divine number.

The Golden Ratio is strongly associated with the **Fibonacci** sequence.

This is a sequence of natural numbers so named after the Italian mathematician **Leonardo Fibonacci** (1170 – 1250) who authored the renowned book "**Liber Apaci**" in 1202 **(Katz 2009)**.

The sequence –as explained in the **Liber Apaci** - cites the population of a number of pairs of rabbits that follow a certain growth rate in which each pair gives birth to a pair of baby rabbits every month. The baby pair gets maturity after a month from birth, starts being productive after two months by giving birth to a pair of rabbits, and continues to do so regularly every month.

Figure 3-11 explains the arithmetic of the growth. Assume that at a given time "r" number of adult and young pairs is: A_r and Y_r respectively. One month after **r**, the Y_r pairs (who were young the previous month) are now adult and will become productive henceforth, so number of adult pairs becomes $A_r + Y_r$. At the same time those who were adult previously (A_r) will give birth to an equal number of young pairs. Two month after "r", rabbits who were young (A_r), are now grown and will add up to the number of adult to become $(A_r + Y_r) + A_r$, while those who were adults previously. $(A_r + Y_r)$ will give birth to an equal number of young pairs. The conclusion depicted from the Fibonacci's growth model presented in figure 3-11 is that the total number of pairs at a given time equals the sum of their numbers in previous two months hence the nth term in Fibonacci sequence equals the sum of the immediately preceding two terms:

$$F_n = F_{n-1} + F_{n-2}$$

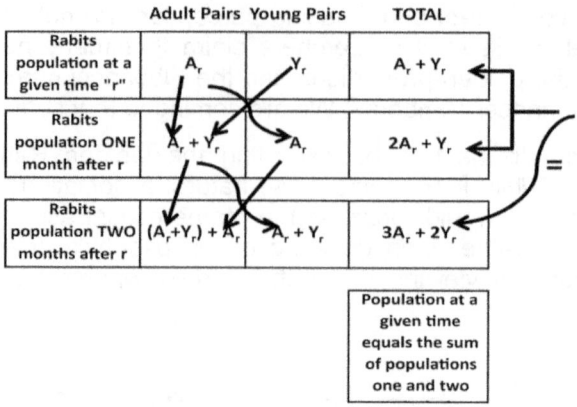

Figure 3-11

Assumptions inherent in this model are that Fibonacci rabbits are immortal, a pair gives birth to a pair of baby rabbits each month and this pair of baby rabbits always comprises a male and a female rabbits who become adults when they are one month of age and breed when they are two month old.

The growth model of Fibonacci rabbits is illustrated in figure 3-12

Starting from the 3rd term, the rth term in the sequence F_r will always be equal to the sum of the two previous terms $F_r = (F_{r-1} + F_{r-2})$, as indicated in in the last column of the tabulation in figure 3-11, and in the growth model of figure 3-12, so Fibonacci sequence will take the form: 1, 1, 2, 3, 5, 8, 13, 21, 34, 55, 89, 144 ... $F_{r-2}, F_{r-1},$

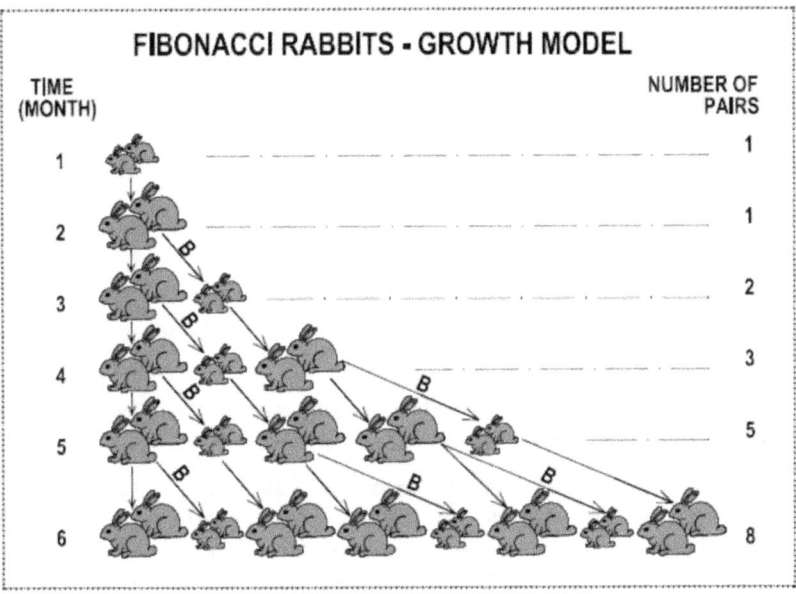

Figure 3-12: Fibonacci's assumptions for the rabbit's multiplication rate

The tabulation of figure 3-13 lists the first 20 terms of Fibonacci sequence, and the ratio F_{r+1}/F_r; that is the ratio (Fibonacci term/previous term)

You will notice that the ratio (F_{r+1}/F_r) converges to φ (The Golden Ratio) as **r** gets larger. This is expressed mathematically as follows:

$$Limit \; (F_{r+1}/F_r) \; \text{as r approaches infinity} \; = \varphi$$

r	1	2	3	4	5	6	7	8	9	10
φ_r	1	1	2	3	5	8	13	21	34	55
$\varphi_{(r+1)}/\varphi_r$	1.0000	2.0000	1.5000	1.6667	1.6000	1.6250	1.6154	1.6190	1.6176	1.6182

r	11	12	13	14	15	16	17	18	19	20
φ_r	89	144	233	377	610	987	1597	2584	4181	6765
$\varphi_{(r+1)}/\varphi_r$	1.6180	1.6181	1.6180	1.6180	1.6180	1.6180	1.6180	1.6180	1.6180	1.6180

Figure 3-13: First 20 terms of Fibonacci sequence.

This amazing result shows the strong association between the **Golden Ratio** and the **Fibonacci** sequence.

However mathematicians are demanding people!. They will not be satisfied with a series that has to be listed sequentially from first term to the term that needs to be evaluated, and they like to know how to calculate the nth term readily, so let us work on that now.

You recall that one of the amazing features of   is that the high powers of it can be reduced to a power of one only, and here are relevant examples:

$$\varphi^2 \qquad\qquad\qquad\qquad = 1 + 1\varphi$$

$$\varphi^3 = \varphi * \varphi^2 = \varphi.(1+\varphi) = \varphi + \varphi^2 \qquad = 1 + 2\varphi$$

$$\varphi^4 = \varphi * \varphi^3 = \varphi.(1+2\varphi) = \varphi + 2\varphi^2 \qquad = 2 + 3\varphi$$

$$\varphi^5 = \varphi * \varphi^4 = \varphi.(2+3\varphi) = 2\varphi + 3\varphi^2 \qquad = 3 + 5\varphi$$

$$\varphi^6 = \varphi * \varphi^5 = \varphi.(3+5\varphi) = 3\varphi + 5\varphi^2 \qquad = 5 + 8\varphi$$

$$\varphi^7 = \varphi * \varphi^6 = \varphi.(5+8\varphi) = 5\varphi + 8\varphi^2 \qquad = 8 + 13\varphi$$

$$\varphi^8 = \varphi * \varphi^7 = \varphi.(8+13\varphi) = 8\varphi + 13\varphi^2 \quad = 13 + 21\varphi$$

You will certainly notice the amazing pattern evident in the numbers before and after the + sign in the right hand part of above equations. These numbers are nothing but the first eight terms of Fibonacci sequence (F_1 through F_8)

From above equations that break down the powers of φ larger than one; we can generalize the rule as follows:

$$\varphi^n = F_{n-1} + F_n *\varphi \qquad ....(1)$$

You will recall that $\varphi = \dfrac{\sqrt{5}+1}{2} = 1.6180339$ is the positive root of the quadratic equation ($\varphi^2 - \varphi - 1 = 0$) and that

$(1 - \varphi) = \dfrac{-\sqrt{5}+1}{2} = -0.6180339$ is the negative root of the same quadratic equation.

Let us now reduce the powers of $(1- \varphi)$ just in the same way we did with the powers of φ before

$(1 - \varphi)$ may also be written as $\qquad\qquad$ **1 - 1φ**

$(1 - \varphi)^2 = 1 - 2\varphi + \varphi^2 \qquad\qquad = 2 - 1\varphi$

$(1 - \varphi)^3 = (1 - \varphi) . (2 - \varphi) = 2 - 3\varphi + \varphi^2 = 3 - 2\varphi$

$(1 - \varphi)^4 = (1 - \varphi)(3 - 2\varphi) = 3 - 5\varphi + 2\varphi^2 \quad = 5 - 3\varphi$

$(1 - \varphi)^5 = (1 - \varphi)(5 - 3\varphi) = 5 - 8\varphi + 3\varphi^2 = 8 - 5\varphi$

$(1 - \varphi)^6 = (1 - \varphi)(8 - 5\varphi) = 8 - 13\varphi + 5\varphi^2 = 13 - 8\varphi$

$(1 - \varphi)^7 = (1 - \varphi)(13 - 8\varphi) = 13 - 21\varphi + 8\varphi^2 = 21 - 13\varphi$

$(1 - \varphi)^8 = (1 - \varphi)(21 - 13\varphi) = 21 - 34\varphi + 13\varphi^2 = 34 - 21\varphi$

And to generalize the power breakdown rule:

$$(1- \varphi)^n = F_{n+1} - F_n \varphi$$

And since $(1 - \varphi) = (-\varphi)^{-1}$, hence:

$$(-\varphi)^{-n} = F_{n+1} - F_n \qquad \ldots\ldots\ldots\ldots\ldots(2)$$

Subtract the two sides of equation (2) from respective sides in equation (1),

$$\varphi^n - (-\varphi)^{-n} = (F_{n-1} + F_n\,\varphi) - (F_{n+1} - F_n\,\varphi),$$

but $F_{n+1} = F_n + F_{n-1}$ (by definition) hence:

$$\varphi^n - (-\varphi)^{-n} = 2\,F_n\,\varphi - F_n = F_n\,(2\varphi - 1) = F_n\,\sqrt{5},$$

hence:

$$F_n = \frac{\varphi^n - (-\varphi)^{-n}}{\sqrt{5}}$$

We have just proved this fantastic formula which defines the nth term of Fibonacci sequence.

This is named **Binet's** formula after the eighteenth century French mathematician **Jacques Binet** (1786 – 1856). It is amazing that irrational numbers like φ raised to a certain power and divided by radical 5, will turn out to be a positive integer F_n .

The formula can also be written in the form:

$$F_n = \frac{\left(\frac{1+\sqrt{5}}{2}\right)^n - \left(\frac{1-\sqrt{5}}{2}\right)^n}{\sqrt{5}}$$

Another analogy that demonstrates the strong association between the Golden Ratio and the Fibonacci sequence is the following analogy in the composition of F_n and φ^n

$$\varphi^n = \varphi^{n-1} + \varphi^{n-2}$$, which was discussed earlier, and:

$F_n = F_{n-1} + F_{n-2}$, which spells out the definition of the nth term

of Fibonacci sequence.

Recall the golden rectangle in figure 3-2 which was split into a square and a smaller golden rectangle on which the same process is performed repeatedly with a golden spiral created in process.

A rectangle proportioned by the Fibonacci numbers 34 and 21 is shown in figure 3-14, and is also split into a square of the dimensions 21x21 and a smaller rectangle of the Fibonacci dimensions 21x13. The splitting process is repeated and a Fibonacci spiral is created.

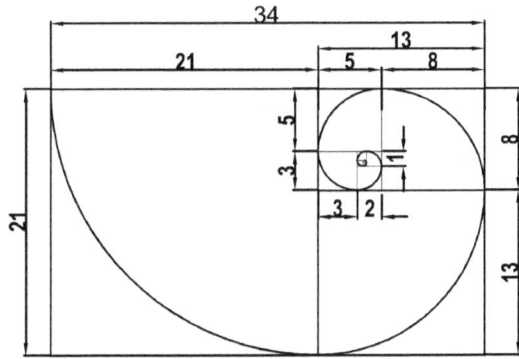

Figure 3-14

The story of Fibonacci's cites a reproduction pattern of rabbits. However hypothetical this pattern is, it remains natural and mimics a natural phenomenon to some extent. That is why many researchers claim that the Fibonacci sequence is often represented in nature.

Like a couple of rabbit gets matured for reproduction two months after birth, a tree branch will start growing from its trunk when the latter is matured enough at age of –say- two units of time. A sub-branch will start growing from a bud on the branch at the same maturity age. A reproduction process mimicking that of the rabbits will go on and on.

No wonder therefore that number of branches of tree counted at different height intervals will follow Fibonacci sequence in the manner shown in figure 3-15

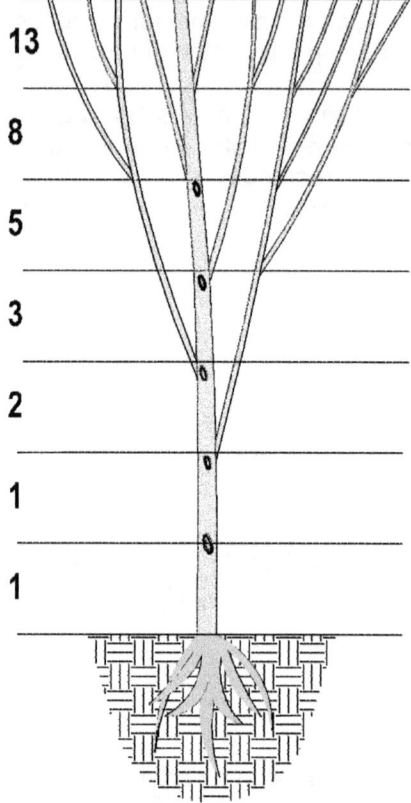

Figure 3-15: Fibonacci sequence in nature

Mathematicians also cite another natural phenomenon flavored with Fibonacci, shown in the amazing visual similarity between the nautilus shell and the Fibonacci spiral (figure 3-16)

In this context I would briefly refer to controversial claims that petals in flowers frequently (and not exclusively) take the numbers 3, 5, 8 or 13, all of which are Fibonacci numbers.

No wonder that most of mathematicians of antiquity have also been philosophers!

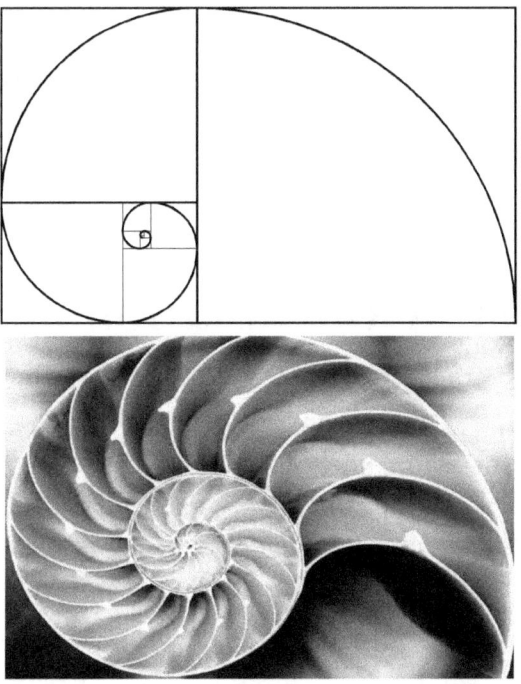

Figure 3-16: Similarity between the nautilus shell and Fibonacci Spiral

The Chapter Quiz

Explain how to construct a pentagram to be inscribed in a circle as to have a given point "a" on the circle circumference as one of its five pointed vertices, using unmarked straight edge and compass.

Discussion and solution of the Chapter Quiz

To construct a pentagram to be inscribed in a circle as to have a given point "a" on the circumference as one of its five pointed vertices; follow the 13 steps listed below:

1. Draw a line from point "a" to the circle center and extend it to meet the circle circumference at point "n" as shown in figure 3-17
2. From the center of the circle draw a radius perpendicular to an. Bisect the radius at point "m"
3. Join the points "n" & "m"
4. Pin the compass at point "m" and draw the arc "4" (figure 3-17)
5. Pin the compass at point "n" and draw the arc "5" to intersect the circumference at point "c"
   Point "**c**" is in fact another pointed vertex of the pentagram
6. Join line segment **ca**
7. Pin the compass at n and draw an arc between point c and d; (its mirror image taking an as a mirror)
8. Join da.
9. Pin the compass at **c** and draw the arc **af** that meets the circumference at **f** which is a third pointed vertex of the pentagram
10. Join **fc**
11. Repeat step 9 but by pinning the compass at **d** and drawing the arc **ab** to locate point **b** (fifth pointed vertex)
12. Join **bd**

13. Join **bf**

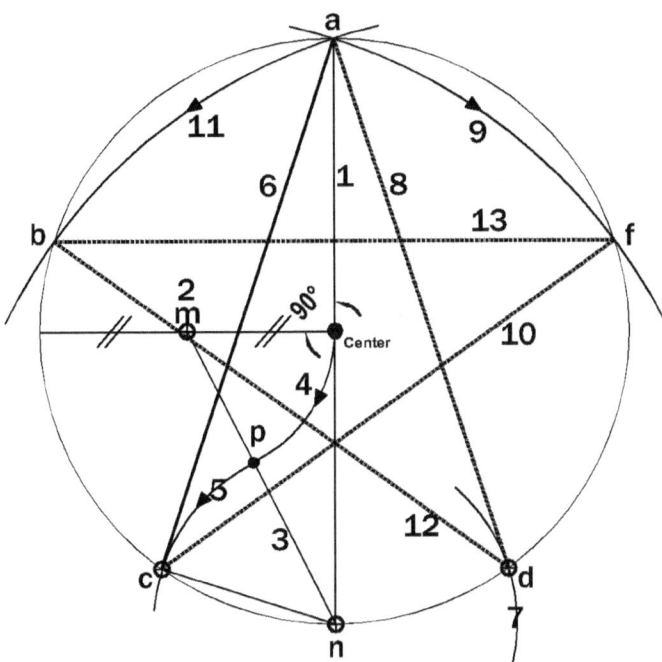

The proof:

Let us assume that the radius of the circle = 2

The angle "**cad**" in the golden triangle in figure 3-18 = **36°** hence angle "**nac**" = $\frac{1}{2}$ **cad** = **18°**.

$$\textbf{sin 18}° = \frac{\textbf{cn}}{\textbf{an}} = \frac{\sqrt{5}-1}{4}$$

Back to figure 3-17:

$$\textbf{mn} = \sqrt{1^2 + 2^2} = \sqrt{5}$$

Arc **5** cuts 1 unit from **mn** hence

pn (the remaining part of **mn** which equals the straight distance **cn**) $= \sqrt{5} - 1$

The **sine** of angle **nac** in figure 3-18 (which is half the pentagram $\dfrac{cn}{an} = \dfrac{\sqrt{5}-1}{4}$ pointed angle) =

hence the angle **nac** = **18°**

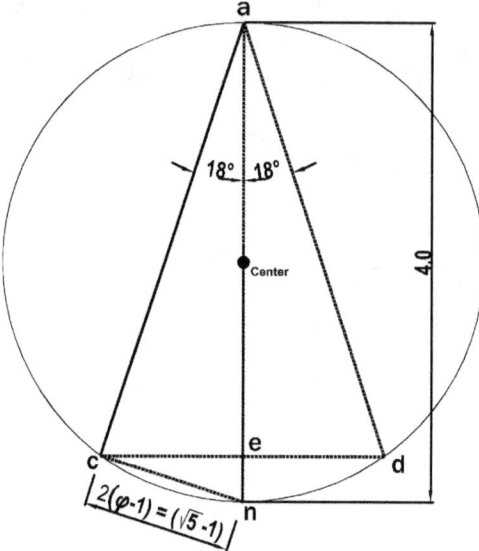

Figure 3-18

Therefore angle cad = **2 * 18 = 36°**; which is the angle of pointed pentagram vertex.

Chapter 4:   The Platonic Solids

In the realm of 2-D geometry, kinds of regular polygons are as infinite as the natural numbers. Each natural number- starting from 3 – can be the number of sides in a regular polygon.

Figure 4-2 shows the first four of these polygons, the equilateral triangle, the square, the pentagon and the hexagon.

An interior angle in a regular polygon would have the measure "**a**"; where

a = 180 (1 – 2/p) - or **π (1 – 2/p)** in radians - where **p** is the number of sides in a polygon.

Consider the regular polygon ABCDE … having "**p**" sides which is shown in figure 4-1.

O is an arbitrary point taken inside the polygon. Join **OA, OB, OC, OD, OE** … to split the regular polygon into "**p**" triangles. Sum of interior angles in each triangle is obviously **180°**, hence the sum of interior angles of the "**p**" triangles is **(180* p)°**

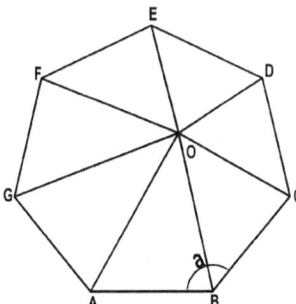

Figure 4-1

Sum of interior angles of the "**p**" triangles = the sum of interior angles of the regular polygon plus **360°** which is the sum of the angles meeting at "**O**", hence:

Sum of interior angles of the regular polygon

= a * p = (180 * p) – 2 * 180 = 180 * (p – 2), therefore

a = 180* (p – 2) / p = 180 (1 – 2/p) = π (1 – 2/p) in radians

For example, the measure of the interior angle in a regular pentagon (**n = 5**) is **180 (1 – 2/5) = 108°**

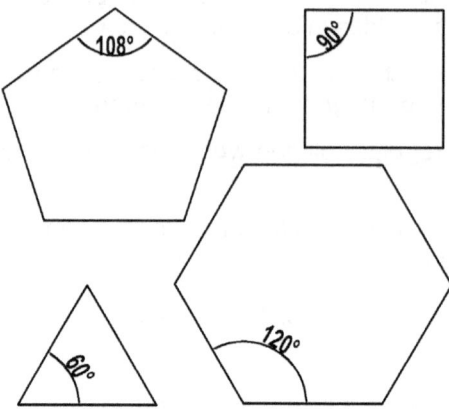

Figure 4-2

A regular polyhedron is a 3-D body formed of a number of regular polygons symmetrically and uniformly positioned similar to the way in which a 2-D regular polygon is formed of line segments of equal measure, uniformly arranged in a polar array.

The discussion here is limited to convex polyhedra which have their faces closing in space

Regular polyhedra are also called Platonic solids.

The hexahedron (the cube) is an example of a polyhedron that we are familiar with.

A logical question may be raised in this context:

Do we have an infinite possible number of regular polyhedra as we do in the case of regular polygons?

The answer – which may be surprising to some of us – is that **we have only five kinds of convex regular polyhedra**.

Let us explore further this piece of information.

There are basic common features for polyhedra; whether or not regular. They are composed of faces each of which is a polygon.

Each two faces meet at an edge and each group of edges meet at a vertex.

In a regular polyhedron, polygons forming its faces are regular and of the same kind.

For example, there is no mix between pentagons and hexagons in the same regular polyhedron (we will see later that such mix exists in a kind of 3-D solids called Archimedean polyhedra). In regular polyhedra, number of edges meeting at a vertex is the same throughout and number of faces meeting at a vertex is also the same at all vertices.

A regular polyhedron can be inscribed by a sphere and can also be circumscribed by another sphere, both of which share a single center called the polyhedron in-center.

There are two essential conditions for a polyhedron (whether or not uniform) to be formed:

- Number of faces meeting at vertex should not be less than **3**

- Sum of interior angles of regular polygons meeting at a vertex should be less than 360o, otherwise; the surface will not close in space (will remain flat).

These two conditions disqualify the regular hexagon or other regular polygons having more than 5 sides from being a face in a regular polyhedron.

The interior angle in a regular hexagon is **120°**, and since a minimum of three faces should meet at a vertex (former condition), sum of angles meeting at each vertex will be **360°** (figure 4-3).

Interior angle at regular polygons with more than **6** sides will even be greater than **120°** hence they cannot be used as faces in a regular polyhedron.

Regular polygons that can be used as faces in a regular polyhedron are strictly the **equilateral triangle**, the **square** and the **pentagon**.

Equilateral triangles may be used as faces in a regular polyhedron if **3**, **4** or **five** of them meet at each of the polyhedron vertices (figure 4-4). If three triangles meet at each vertex a **Tetrahedron** – composed of 4 triangular faces will be resulted.

In case four triangles meet at each vertex the resulting polyhedron will be the **Octahedron** which has eight faces.

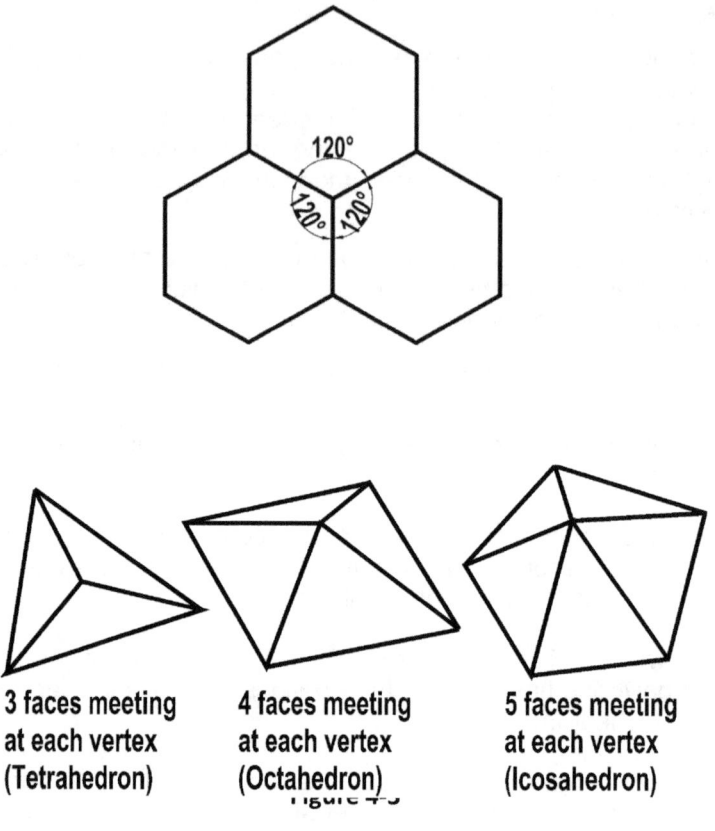

3 faces meeting at each vertex (Tetrahedron)　　**4 faces meeting at each vertex (Octahedron)**　　**5 faces meeting at each vertex (Icosahedron)**

Figure 4-4

When 5 equilateral triangles meet at each vertex the resulting polyhedron will be the icosahedron which will be discussed shortly. Squares are used as faces in the **Hexahedron** (the cube), in which three faces meet at each vertex.

The fifth and last regular polyhedron is the **Dodecahedron** which is made up of 12 faces of regular pentagons, with each three faces meeting at a vertex of the polyhedron.

The tabulation in figure 4-5 lists the five regular polyhedron and information on the regular polygon that forms each face in the polyhedron.

I shall be discussing here the composition and structure of each of the five regular polyhedra.

A part of the discussion will focus on the geometric properties of the polyhedron in which the length of the edge is assumed to measure 1.0 linear unit.

THE FIVE REGULAR POLYHEDRA (PLATONIC SOLIDS)				
POLYHEDRON	FACE POLYGON .	NUMBER OF POLYGONS AT EACH VERTEX	INTERIOR ANGLE OF THE POLYGON	SUM OF ANGLES AT A VERTEX
Tetrahedron	E. L. Ttriangle	3	60	180
Octahedron	Eq L Ttriangle	4	60	240
Icosahedron	Eq L Ttriangle	5	60	300
Hexahedron	Square	3	90	270
Dodecahedron	Pentagon	3	108	324

Figure 4-5

1 -The Tetrahedron

The tetrahedron (figures 4-6a) takes the shape of a pyramid having its base and other three faces formed of equilateral triangles.

Geometric description - based on length of 1.0 unit for the edge - is provided in figure 4-6b

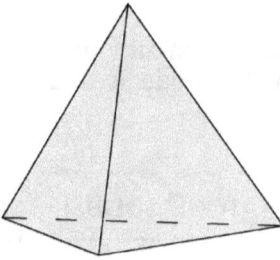

Figure 4-6a: The Tetrahedron

Number of faces   **4**

Number of edges   **6**

Number of vertices   **4**

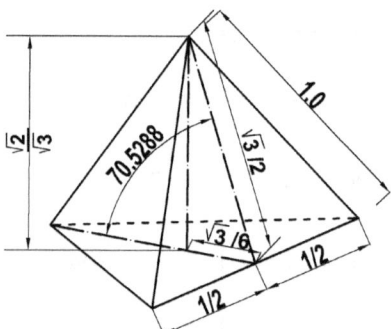

Figure 4-6b: The Tetrahedron - Geometric description

2 - The Octahedron

The octahedron (figure 4-7a) takes the shape of 2 pyramids placed opposite to each other at the square base with all its 8 faces formed of equilateral triangles. Assembly two-halves image is shown in figure 4-7b

Number of faces   **8**

Number of edges   **12**

Number of vertices   **6**

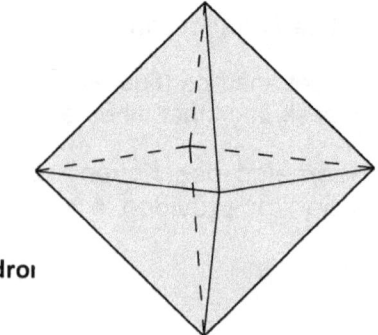

Figure 4-7a: The Octahedroı

Figure 4-7b: The Octahedron disassembelled

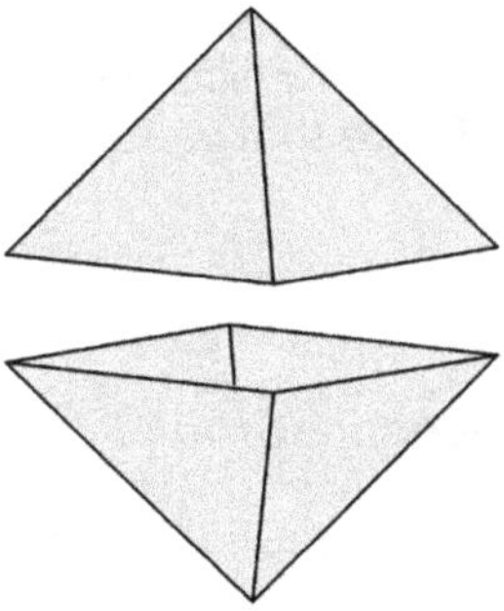

Figure 4-7c: The Octahedron. Geometric description based on edge length of 1.0 unit

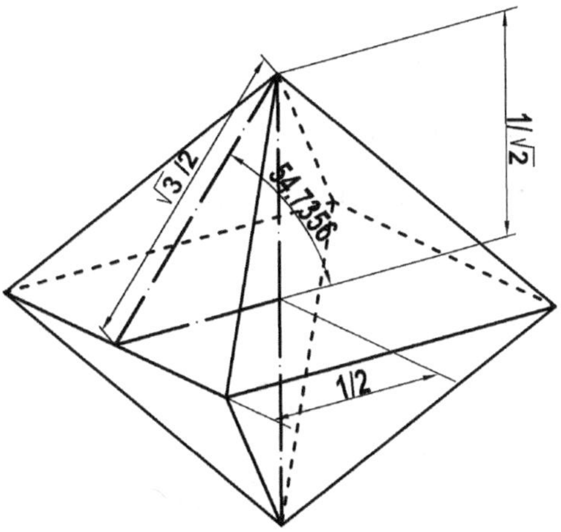

3- The Hexahedron (the Cube)

The hexahedron (figure 4-8) is simply the cube, which is made up of six squared faces.

Number of faces   **6**

Number of edges   **12**

Number of vertices   **8**

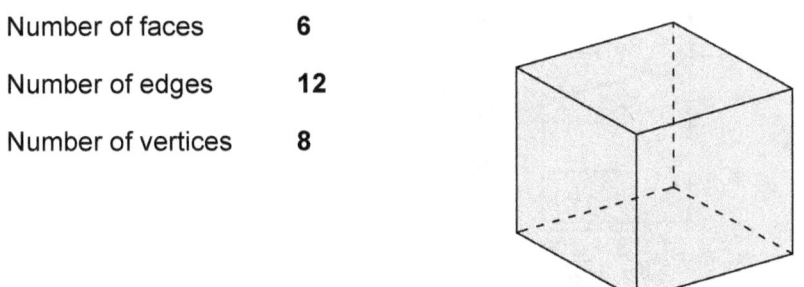

Figure 4-8: The Hexahedron (the cube)

Since we are familiar with the hexahedron and since it is pretty simple, I figure that we do not need any geometric description for it.

4 – The Icosahedron

The icosahedron (figure 4-9a) is formed of 20 faces of equilateral triangles.

Number of faces **20**

Number of edges **30**

Number of vertices **12**

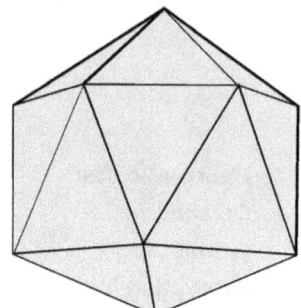

Figure 4-9a: The

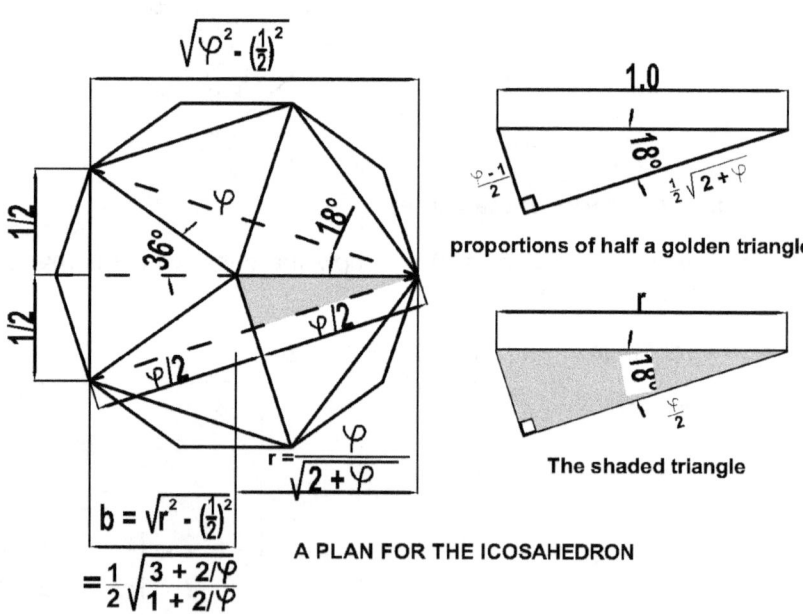

Figure 4-9b: The Icosahedron – Geometric Description I

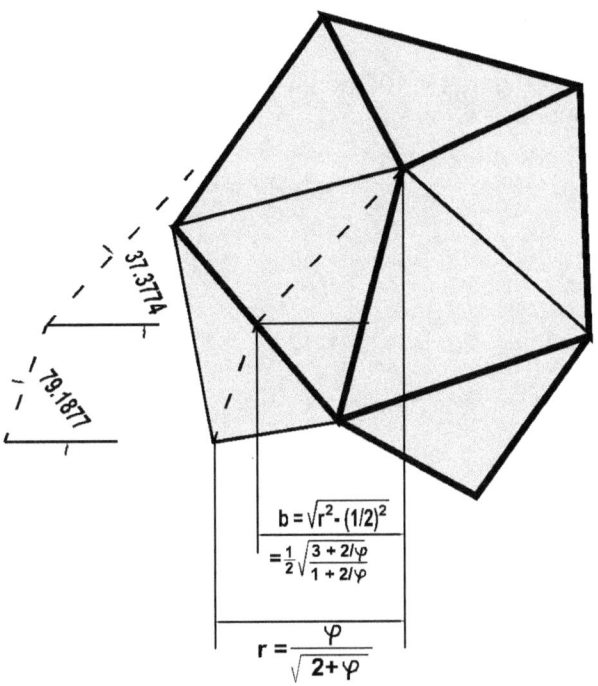

Figure 4-9c: The Icosahedron – Geometric Description II

The Greek letter φ (pronounced "phi") that appears in several dimensions of the icosahedron (and will also appear in geometric descriptions of the dodecahedron) is the Golden Ratio:

$$\varphi = \frac{\sqrt{5}+1}{2} \quad .$$

We have seen in Chapter 3 how far these two polyhedra are so associated with the Golden Ratio φ

5 – The Dodecahedron

The dodecahedron (figures 4-10a and 4-10b) is formed of 12 faces of uniform pentagons.

Number of faces   12

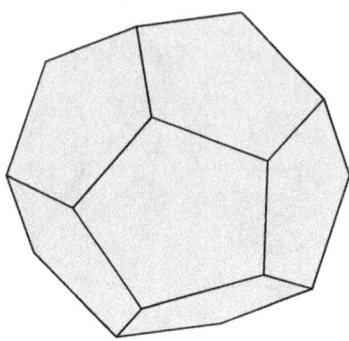

Number of edges   30

Number of vertices   20

Figure 4-10a: The Dodecahedron

Figure 4-10b:

The Dodecahedron (upper half) - Geometric Description based on edge length of 1.0 unit

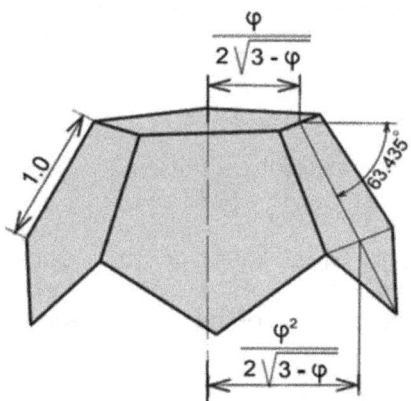

The tabulation in figure 4-11 shows the structural composition of the five regular polyhedra, that is number of faces, vertices and edges.

It is noticeable that the sum of number of faces and number of vertices in any of the five regular polyhedra equals the number of edges plus 2 (last shaded two columns in the tabulation), so:

F + V = E + 2.

It would be interesting to find out whether such a relationship that governs the number of faces, vertices and edges is specific to our five regular polyhedra only, or it is a global rule that applies on all 3-D solids.

This question is dealt with by a branch of geometry called **Topology**.

REGULAR POLYHEDRA - STRUCTURAL COMPOSITION						
POLYHEDRON	FACE POLYGON .	NUM. OF FACES "F"	NUM. OF VERTICES "V"	NUM. OF EDGES "E"	F + V	E + 2
Tetrahedron	Equilate. Ttriangle	4	4	6	8	8
Octahedron	Equilate. Ttriangle	8	6	12	14	14
Icosahedron	Equilate. Ttriangle	20	12	30	32	32
Hexahedron	Square	6	8	12	14	14
Dodecahedron	Pentagon	12	20	30	32	32

Figure 4-11: Structural composition of Regular Polyhedra

Topology is concerned with non-dimensional geometric properties and configuration aspects that are not affected by elastic deformation such as stretching or twisting of the object's body or surface.

Consider any polyhedron - not necessarily regular - as that shown in figure **4-12**.
We have to agree on a suitable name for it so let us call it *"anyhedron"*

In order for us to be able to study its topological properties we would convert it to a 2-D object.

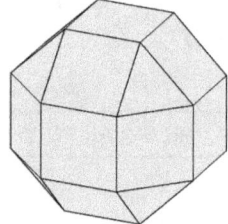

Figure 4-12: Anyhedron

Because we are three dimensional beings; we can view two dimensional objects from outside and can easily have insight through their simpler properties way easier than we do with our pears of three dimensions.

(We shall see in chapter 10 even more difficult monster objects that have four dimensions)

But how shall we downgrade our anyhedron to a two dimensional object?

Let us first unfold it to visualize its properties. The unfolded shape in figure 4-13 reveals that there are **26** faces in our Anyhedron, **8** of them are equilateral triangles and the remaining are squares.

But other than seeing its faces in undistorted dimensions; unfolding the Anyhedron will not help us identifying the topographical relationship between the number of faces, number of vertices and that of the edges. We need to use the common topographical tool of stretching the Anyhedron flat without separating its faces from one another.

Figure 4-13: The Anyhedron unfolded

But how shall we do that?

Here is the recipe:

- Remove one of its faces – say- that squared one at its top level to create an opening through its interior,
- Then - imagining it is made of thin elastic rubber – stretch the edges of that removed top face that became an opening all the way until the whole rubber surface once was 3-D object is turned out to be a plane surface as that shown in figure 4-14.

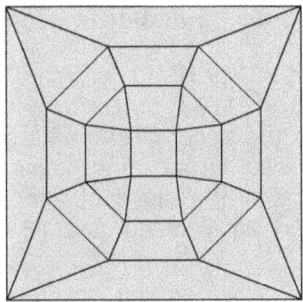

Figure 4-14: The Anyhedron stretched out plane

Do not get concerned that the size of its faces or edges will be distorted big time. This is not an issue because topology is not concerned with any dimensions as stated earlier.
What next?

We would gradually strip the plane stretched rubber of our Anyhedron from its distorted faces – not squares or equilateral triangles any further – and carefully monitor how would the expression "**F + V - E**" (that is number of faces + number of vertices – number of edges) be affected every time we remove faces from our stretched Anyhedron.

The story is explained in 8 consequent steps in figure 4-15.

It tells us that if the plane Anyhedron is stripped of all its faces except that squared one at the bottom, **F + V - E** will remain unchanged. We know that there are 4 vertices and 4 edges in this single squared face of the Anyhedron, for which $(F + V - E) = 1 + 4 - 4 = 1$

And since the stripping process did not change this relationship, it is deduced that the same relationship $(F + V - E) = 1$ will also apply to the stretched rubber shape of figure 4-15. We remember that at the beginning of this analysis; we had to remove the top squared face in the Anyhedron to be able to stretch it plane from its free edges. Above relationship should therefore be corrected for 3-D polyhedra objects by adding 1 to its right side. It will read:

$F + V - E = 2$, or $F + V = E + 2$

(**Number of faces + number of vertices = number of edges + 2**).

Historically this formula was first proved by the Swiss mathematician Leonhard Euler (1707 – 1783) that is why it bears his name. We will know more about the accomplishments of Euler in chapter 6.   Euler's formula is valid for all three dimensional polyhedra solids in general and not only in regular polyhedra.   In our Anyhedron, there are **26** faces, **24** vertices and **48** edges, hence **$F + V = 50 = E + 2$.**

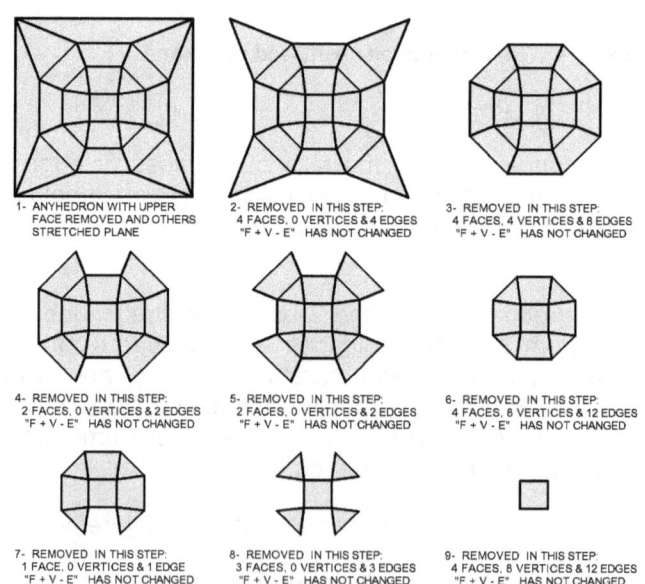

1- ANYHEDRON WITH UPPER FACE REMOVED AND OTHERS STRETCHED PLANE

2- REMOVED IN THIS STEP: 4 FACES, 0 VERTICES & 4 EDGES "F + V - E" HAS NOT CHANGED

3- REMOVED IN THIS STEP: 4 FACES, 4 VERTICES & 8 EDGES "F + V - E" HAS NOT CHANGED

4- REMOVED IN THIS STEP: 2 FACES, 0 VERTICES & 2 EDGES "F + V - E" HAS NOT CHANGED

5- REMOVED IN THIS STEP: 2 FACES, 0 VERTICES & 2 EDGES "F + V - E" HAS NOT CHANGED

6- REMOVED IN THIS STEP: 4 FACES, 8 VERTICES & 12 EDGES "F + V - E" HAS NOT CHANGED

7- REMOVED IN THIS STEP: 1 FACE, 0 VERTICES & 1 EDGE "F + V - E" HAS NOT CHANGED

8- REMOVED IN THIS STEP: 3 FACES, 0 VERTICES & 3 EDGES "F + V - E" HAS NOT CHANGED

9- REMOVED IN THIS STEP: 4 FACES, 8 VERTICES & 12 EDGES "F + V - E" HAS NOT CHANGED

Figure 4-15: Eight steps to prove Euler theorem: $F + V = E + 2$

I must however highlight as a condition for the validity of Euler's formula that it will not work if there is a hole in the polyhedron like that one created in our Anyhedron. (figure 4-16)

Two holes were made in our Anyhedron (see figure 4-16) which resulted in the addition of **10** new faces, **16** vertices and **24** edges. The count of **F + V - E** has therefore increased by 10 + 16 – 24 = 2, which would invalidate Euler's formula.

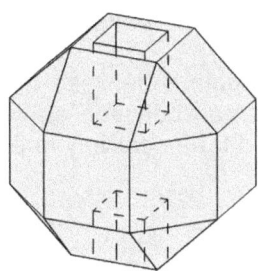

Figure 4-16: A polyhedron with two holes

DIHEDRAL ANGLE

The interior angle between any two adjoining faces of a polyhedron is called the *dihedral angle*. In a regular polyhedron the dihedral angle is fixed; and is one of its prominent properties.

Dihedral angle α in a polyhedron is a function of the following two variables:

- Number of faces meeting at a vertex "q". This can be 3, 4 or 5
- Number of side in the face polygon "p". This also can be 3, 4 or 5

Consider the polyhedron vertex "v" in figure 4-17a at which edges **va**, **vb**, **vc** and **vd** meet.

Although number of faces "q" meeting at vertex **v** is shown in the figure to be 4, the figure should be seen to represent the general condition applicable for **q = 3**, **4** or **5**.

As stated in the first paragraph of this chapter, the interior angle in a regular polygon would have the measure $\pi\ (1 - 2/p)$, hence angle **avb** in figure 4-17a should measure $\pi\ (1 - 2/p)$.

Obviously, this is also the measure of angles **bvc**, **cvd** and **dva**.

It follows that the base angle **vab** (and its pears) measure

$(\pi - \pi\ (1 - 2/p))/2 = \pi/p$.

Since **av** (the face polygon side length) is assumed to be **1.0** linear unit, the measure of **ab** is $2\ cos\ (\pi/p)$ as shown in figure 4-17a.

Dihedral angle α between the two adjacent faces **vad** and **vab** is the angle **deb** between **ed** and **eb**; which are the perpendiculars from **d** and **b** respectively to the common edge **av**.

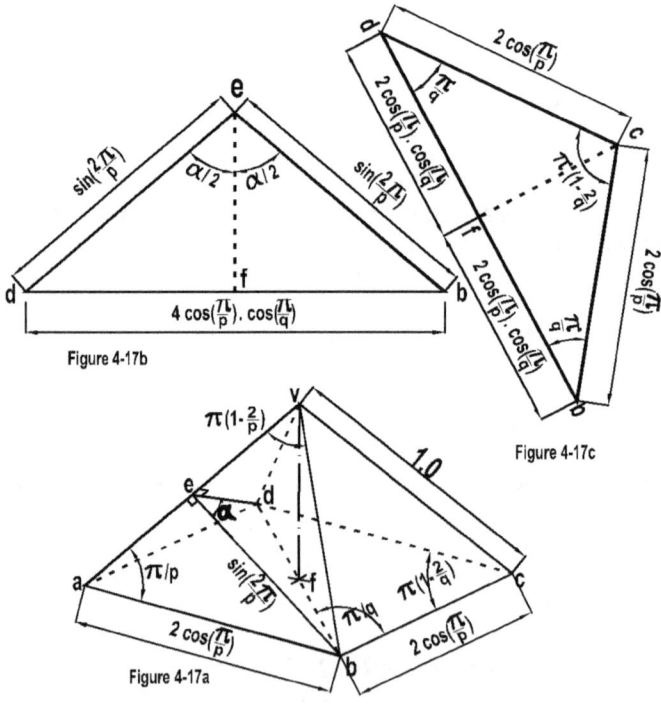

Figure 4-17b

Figure 4-17c

Figure 4-17a

Figures 4-17a , 4-17b and 4-17c

To be able to focus on the isosceles triangle **deb** that contains the angle α between its equal sides. It is separately drawn in figure 4-17b. You may not recognize it at first glance due to its 3D distorted shape in figure 4-17a.

To find the measure of **db** (the hypotenuse of the isosceles triangle in figure 4-17c); the adjacent triangle **dbc** (which is also isosceles) is drawn separately in the 2D undistorted shape of figure 4-17c.

From the triangle in figure 4-17c:

$$\mathbf{df} = \mathbf{fb} = 2\cos\left(\frac{\pi}{p}\right)\cos\left(\frac{\pi}{q}\right), \text{ hence}$$

$$\mathbf{db} = 4\cos\left(\frac{\pi}{p}\right)\cos\left(\frac{\pi}{q}\right)$$

This measure of the hypotenuse **db** is brought back to figure 4-17b.

From figure 4-17b:

$$\sin\left(\frac{\alpha}{2}\right) = \left(2\cos\left(\frac{\pi}{p}\right).\cos\left(\frac{\pi}{q}\right)\right)/\left(\sin\left(\frac{2\pi}{p}\right)\right)$$

$$\sin\left(\frac{2\pi}{p}\right) = 2\sin\left(\frac{\pi}{p}\right).\cos\left(\frac{\pi}{p}\right)$$

Since and substituting back in the above equation we get this beautiful formula for the dihedral angle α – between two adjoining faces:

$$\sin\left(\frac{2\pi}{p}\right) = 2\sin\left(\frac{\pi}{p}\right).\cos\left(\frac{\pi}{p}\right)$$

The tabulation in figure 4-18 lists the five platonic solids and the dihedral angle of each of them calculated according the above formula.

POLYHEDRON		NUMBER OF EDGES IN EACH FACE p	NUMBER OF FACES MEETING AT A VERTEX q	DIHEDRAL ANGLE (in degrees) α
Tetrahedron	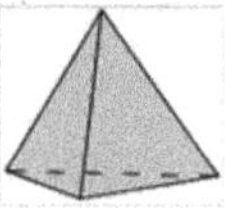	3	3	70.53
Octahedron	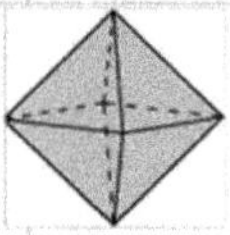	3	4	109.47
Icosahedron	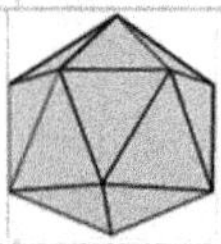	3	5	138.19
Hexahedron	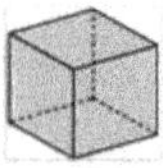	4	3	90.00
Dodecahedron	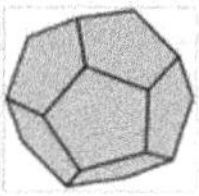	5	3	116.57

Figure 4-18

ARCHIMEDEAN POLYHEDRA

There is a category of polyhedra which are not as regular as the platonic solids but still have a lot of its symmetry.

These are called **Truncated** or **Archimedean Polyhedra**. The difference between regular and regular polyhedra is that a truncated polyhedron can combine two or more kinds of polygons, but it can still be inscribed in a sphere and can be circumscribed by a sphere.

Symmetry in the composition of vertices will be maintained in truncated polyhedra in the sense that the same combination of different polygonal faces will be meeting at each vertex.

A popular example of truncated polyhedra is that which shapes any football. This is called "truncated icosahedron" (figures 4-19a and 4-19b)

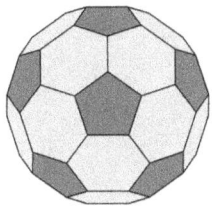

Figure 4-19a: Truncated Icosahedron

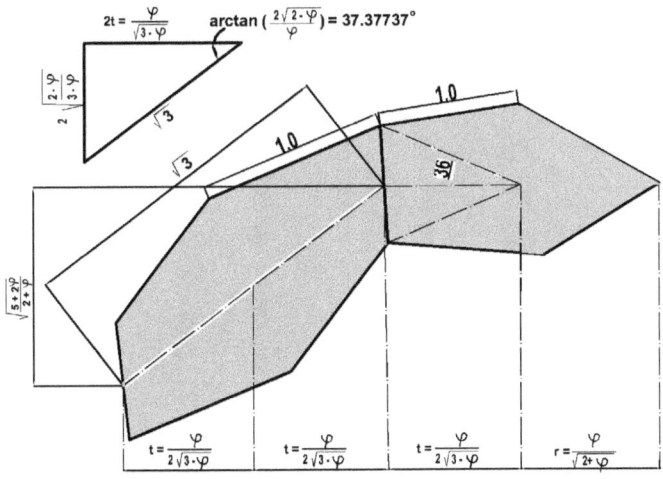

$$2t = \frac{\varphi}{\sqrt{3 \cdot \varphi}} \qquad \arctan\left(\frac{2\sqrt{2 \cdot \varphi}}{\varphi}\right) = 37.37737°$$

$$t = \frac{\varphi}{2\sqrt{3 \cdot \varphi}} \qquad t = \frac{\varphi}{2\sqrt{3 \cdot \varphi}} \qquad t = \frac{\varphi}{2\sqrt{3 \cdot \varphi}} \qquad r = \frac{\varphi}{\sqrt{2 + \varphi}}$$

Figure 4-19b: Truncated Icosahedron - Geometric description based on edge length of 1.0 unit

And you know what: You remember our Anyhedron (figure 4-12). This is also a truncated polyhedron. It is formed of 8 equilateral triangles and 18 squares, and it is called: *Pseudo Rhombicubotahedron*. I know that you prefer the name I gave.

The Chapter Quiz

Duality in the geometry of regular polyhedra means that the center points of the polygonal faces in one polyhedron are the vertices of another smaller one, which might also (and might not) be of the same kind.

In the five Platonic solids, you may join the center points of all the polygonal faces with line segments such that these line segments become the edges of another polyhedron.

Show graphically the phenomenon of duality in the five Platonic solids.

Discussion and solution of the Chapter Quiz

Starting with the tetrahedron, and keeping in mind that number of faces in it is 4, we would logically seek a dual polyhedron having only 4 vertices.

The tetrahedron is the only polyhedron which has the number of faces equaling the number of vertices ($F = V = 4$).

Accordingly, the tetrahedron is said to be **self-dual** as shown in figure 4-20

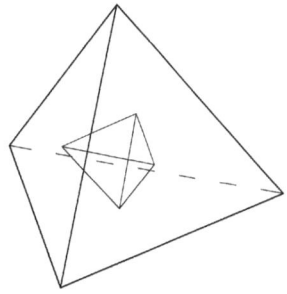

Figure 4-20: The Tetrahedron is "self-dual"

The hexahedron and octahedron are **mutually dual**. That is a hexahedron is generated by joining the center-points of the faces of an octahedron (figure 4-21a) and an octahedron is generated by joining the center-points of faces of a hexahedron (figure 4-21b).

Such mutual duality is enabled by the mutual compatibility of number of faces and number of vertices in the hexahedron and the octahedron:

Number of faces in an octahedron = number of vertices in a hexahedron = 8, and

Number of faces in a hexahedron = number of vertices in an octahedron = 6

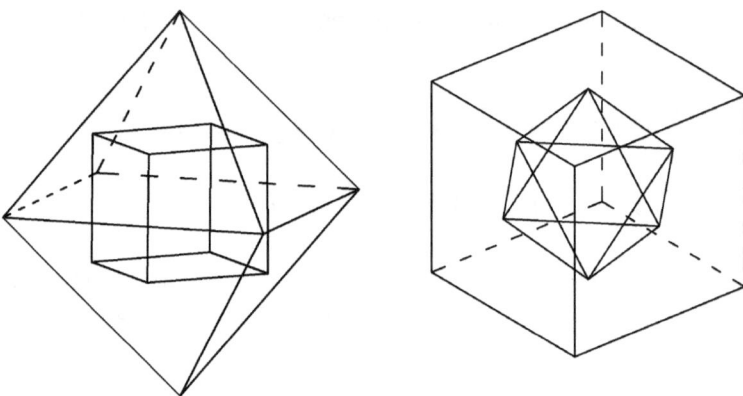

Figure 4-21a:	Figure 4-21 b:
Octahedron	Hexahedron
enveloping	enveloping
Hexahedron	Octahedron

In the same way; the icosahedron and the dodecahedron are *mutually dual*. That is a dodecahedron is generated by joining the center-points of the faces of an icosahedron (figure 4-22a) and an icosahedron is generated by joining the center-points of the faces of a dodecahedron (figure 4-22b).

Such mutual duality is enabled by mutual-compatibility of number of faces and vertices in the icosahedron and the dodecahedron:

Number of faces in an icosahedron = number of vertices in a dodecahedron = 20, and

Number of faces in a dodecahedron = number of vertices in an icosahedron = 12

Mutual duality between the octahedron and the hexahedron

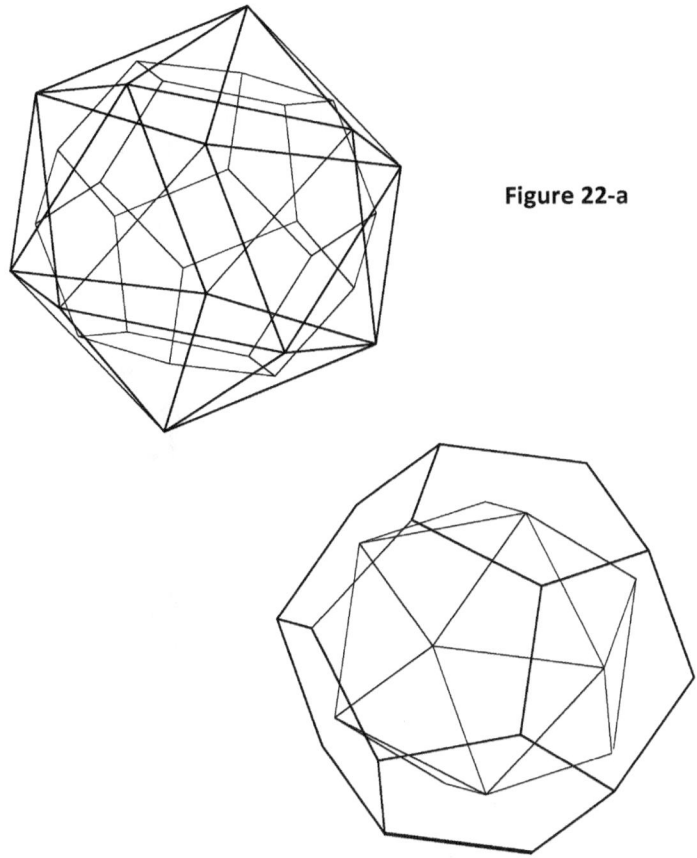

Figure 22-a

Figure 4-22b: Mutual duality between the Icosahedron and the Dodecahedron

Chapter 5: Modular Arithmetic

WHAT IS MODULAR ARITHMETIC

Modular Arithmetic systems deal with whole numbers and starts counting from 1 up to a certain number (called the modulus). After reaching the modulus counting starts over again in a cyclic pattern from the beginning to the modulus

Modular Arithmetic was introduced in its modern style in 1801 by **Carl Friedrich Gauss** (1777-1855).

Figure 5-1: Carl Friedrich Gauss

Modular arithmetic is actually in continual use by all of us in counting the months of a year, days of a week, hours of a clock, minutes in an hour and seconds in a minute. In the case of counting the months of a year – for example - counting starts with one and goes up to 12 (the modulus in this case) then starts at one again in another round, and keeps repeating the cycle endlessly.

I should however mention that Gauss suggested that modular counting would start from 0.

For example the hours of a day would be 0, 1, 2 .. 11 and then count back to 0 again.

Basic arithmetic operations are performed differently in modular arithmetic.

For example if you want to add **10** hours to **7** o'clock it will be **5** o'clock so **7 + 10 = 5** in this case, but you would use the term **congruent** in lieu of **equal**.

To express such a nodular congruence; Gauss introduced the sign $\equiv$ in lieu of the sign =, hence a rather meaningful way to write addition expression in our example is: **7 + 10 $\equiv$ 5**

DIVISIBILITY AND REMAINDERS

Modular Arithmetic also deals with the divisibility of whole numbers by one another, and by finding the remainder in any division.

For example; if you want to express the number of weeks existing in 60 days; you say that 60 days are equivalent to 8 weeks and 4 days. This is also similar to saying: 60/7 = 8 plus a remainder of 4. As per the notation of Modular Arithmetic; the statement is expressed as:

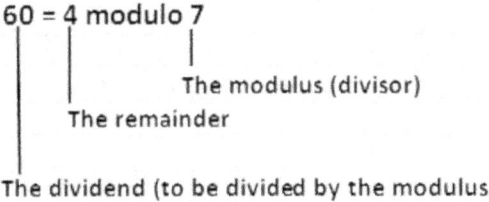

The modulus (divisor)

The remainder

The dividend (to be divided by the modulus)

I shall be using the abbreviation "**mod**" in lieu of "**modulo**" from this point in this chapter.

"**Sun Tsu**" (544 – 496 BC) was a Chinese military general, strategist and philosopher of ancient time. In this context; he is remembered for the following mathematical problem – related to modular arithmetic- that was posed in his book: "**Sun Tsu** manual":

Certain objects of unknown number were grouped in different ways:

If grouped in packs of three pieces each; the remainder will be 2

If grouped in packs of five pieces each; the remainder will be 3

If grouped in packs of seven pieces each; the remainder will be 2.

Find the least possible number of objects.

I shall express below the information provided in this puzzle in the just discussed modular notation; assuming the number of objects is "**n**"

"If "n" is grouped in packs of three pieces each; the remainder will be 2":   n = 2 mod 3

..........(1)

"If "n" is grouped in packs of five pieces each; the remainder will be 3":   n = 3 mod 5

..........(2)

"If "n" is grouped in packs of seven pieces each; the remainder will be 2":   n = 2 mod 7

..........(3)

Since the remainder in both equations **1** and **3** is **2**, it will be convenient to merge them in a single equation. The basis for that is simple: If the remainder in the division "**n/ a**" is the same remainder in the division "**n/ b**", then the remainder will also be the same in the division "**n/ k**" where "**k**" is the Least Common Multiple of "**a**" and "**b**".

To make it clear I shall give another numerical example using simple arithmetic notation.

Let **n=77**, a = **6** and b = **8**.

We know that if you divide **77÷ 6** the remainder will be **5**

And if you divide **77÷ 8** the remainder will be **5** (also)

Then it follows that if you divide **77÷ LCM (the Least Common Multiple)** of **6** and **8**; the remainder will also be **5**.

Definition: "*The Least Common Multiple of two or more integers is the smallest number that is a common multiple of these integers*"

LCM (the **Least Common Multiple**) for **6** and **8** can be found as follows:

Accordingly, if you divide **77÷ 24 (the LCM of 6 & 8)** the remainder will be **5** (also)

Back to **Sun Tsu** problem:

If you divide **n÷ 3** the remainder will be **2**   **(equation 1)**, and

If you divide **n÷ 7** the remainder will also be **2** **(equation 3),**

Factor	1	2	3	4	5	6
Mutiples of 6	6	12	18	24	30	36
Mutiples of 8	8	16	24	32	40	48

Least Common Multiple for 6 and 8 is 24

and since both divisors (**3** and **7**) are prime numbers, the Least Common Multiple for them is **3** * **7 = 21**

Therefore, if you divide **n ÷ 21** **(LCM of 3 & 7)** the remainder will also be **2**, so we can replace equations **1** & **3** by this equation:

n = 2 mod 21

$$.........(4)$$

which means in simple language that the remainder of the division **n / 21** is **2**, so it can also be expressed in this way:

n = 21 k + 2 .... **(5)**

Similarly, equation **2** may be written in this simpler form:

n = 5 r + 3 .... **(6)**

where "**k**" and "**r**" are positive integers to be determined such hat the conditions of equations **5** and **6** are met.

Solution of equations (**5**) and (**6**) can be reached by trial and error. That is to first assume **k = 1**, get **n** in equation (**5**) then to substitute its value in equation (**6**) and find if such substitution will result in an integer "**r**". If not try k = 2 and repeat the same test. The reader should not be concerned that the trial-error loop might go endlessly. In fact in the example at hand, the maximum number of trials will be **5**, which is the modulus (divisor) in equation **6**.

However, it turned out to be only one trial because if you substitute k=1 in equation (**5**), you get **n = 23**. Now let **n = 23** in equation (**6**), **r** will be **4** which is an integer (luckily), hence the happy solution is **n = 23** reached at first trial.

We should know however that while **23** is the least value for "**n**" that satisfies the three modular conditions of the problem; it is not the only correct solution. You will also get correct answers for **k = 6,**

11, 16, 21, 26, 31.. or in general **k = 1 + 5q** where **q** is any positive integer.

The reader may notice that these valid values of **k** have a common difference of **5** between each two successive values. Why **5**?, you may ask. The answer is: "*because 5 is the modulus in equation 6.*"

THE PROBLEM OF BERMA

This is also a divisibility related problem about a number of objects "**n**" that satisfy multiple modular conditions. It is called the "***Problem of Berma***" which is taken from Egyptian folk heritage. The problem goes as follows:

A villager from a small village called Berma in the Nile Delta was carrying on top of her head a basket of eggs; and was heading to the marketplace to sell them when someone rushing on his bicycle collided with her causing the basket to fall and the eggs to smash.

When she appeared before the judge seeking compensation for the loss she had incurred; the judge asked if she could remember the number of eggs which had been in the basket. She replied:

"*I do not remember your honor the exact number, but I do remember that when I grouped them in packs of two, packs of three, packs of four, packs of five or packs of six each; the remainder was always one egg. However - your honor - when I grouped them in packs of seven each; no leftovers were there*"

It took the judge a few minutes to decide what the minimum number of eggs that would satisfy the multiple modular conditions claimed by the villager was.

Can you envisage what verdict he must have reached at?

In search for a solution; I shall avoid using the formal modular arithmetic terms because I know they do not appeal to many of us. I shall put the information given in the simple form used earlier in equations (5) and (6).

First step in solving the problem is to find the multiples of **2**, **3**, **4**, **5** and **6**, then to depict the Least Common one out of these multiples as done earlier in **Sun Tsu's** problem.

Factor	1	9	10	11	12	13	14	15	16	17	20	30
Mutiples of 2	2	18	20	22	24	26	28	30	32	34	40	60
Mutiples of 3	3	27	30	33	36	39	42	45	48	51	60	90
Mutiples of 4	4	36	40	44	48	52	56	60	64	68	80	120
Mutiples of 5	5	45	50	55	60	65	70	75	80	85	100	150
Mutiples of 6	6	54	60	66	72	78	84	90	96	102	120	180

Least Common Multiple for 2, 3, 4, 5 and 6 is 60

The Least common Multiple for **2, 3, 4, 5** and **6** is therefore **60**.

This means that: **60 k** will be divisible by **2, 3, 4, 5**, or **6** with no leftover remained (**k being any positive integer**). Since the remainder in dividing **n** by **2, 3, 4, 5**, or **6** will always be **1**, we can therefore express **n** in the form:

n = 60 k + 1 ....(*)

60 k + 1 should however be divisible by **7** with o remainder as per the conditions of the problem

To simplify the search effort, we would put the above equation in the form: **n = 56 k + (4k+1)**

We know however that **56 k** is divisible by **7**, hence should find **k** such that **(4k+1)** will be divisible by **7**.

As done in solving **Sun Tsu** problem; we shall try values of **k** that would satisfy such a modular condition, knowing in advance that our trials will never exceed seven.

Soon we shall find that **k=5** will make **(4k+1)=21** ; which is divisible by **7** hence can substitute **k=5** in equation (*) to get **n = 60*5 +1 = 301** as a solution for the **Berma** problem.

Again; I must remind the reader that **301** is the solution corresponding with the least number of eggs, and not the only solution. Naturally, in absence of concrete evidence; the judge will only be able to establish a compensation equivalent to the least possible number of eggs that satisfies all the given modular conditions. Other solutions are attained by adding multiples of **420** to the least solution (the 301 eggs), so **721, 1141, 1561, 1981, 2401, 2821**... etc. are also correct solutions.

But why **420**?

420 is the Least Common Multiple of **2 through 7**. Adding **420** or its multiples to **301** will not change the combined modular conditions given in the problem. This means that the number obtained by any of such additions will still leave a remainder of 1 if divided by 2, 3, 4, 5 or 6, and will still be divisible by 7.

The general solution for the Berma Problem will therefore be:

n = 420 k -119, where **k** is any positive integer starting with **1**.

DIVISIBILITY RULES

In our every day's arithmetic practice we use a decimal numbering system in which the position of a digit represents its value. Starting from right to left; digits represent units, tens, hundreds, thousands and so on. For example, a number like *24,876* has the value of (*6 + 7*10 + 8*100 + 4*1000 + 2*10000*).

I shall use here the notation: D_0, D_1, D_2, D_3, D_4, D_5 ... etc. to denote the digits representing **units**, **tens**, **hundreds**, **thousands** and **ten thousands** respectively of a numeral **N** which has a numerical value $N = D_0 + D_1*10 + D_2*100 + D_3*1000 + D_4*10000$.

In the number *24,876* in the example given earlier; $D_0 = 6$, $D_1 = 7$, $D_2 = 8$, $D_3 = 4$ and $D_4 = 2$

From the equation ($N = D_0 + D_1*10 + D_2*100 + D_3*1000 + D_4*10000$); we logically establish that the divisibility of **N** (we call it the **dividend**) by another number **DV** (we call it the **divisor** or **modulus**) is determined by the divisibility of **10, 100, 1000, 10000** etc. by **DV.**

I shall present here simple methods to examine the divisibility of a numeral **N** by a divisor **DV = 2, 3, 5, 7, 8, 9, 11** or **13**, that is to establish whether a number like **4653** (for example) is divisible by **11** without a remainder. There is a specific Divisibility Testing method for each one of these divisors and each method will be validated by applying the general procedure outlined below.

Suppose you want to validate the testing method suggested to establish whether a dividend $N = D_0 + D_1*10 + D_2*100 + D_3*1000 + D_4*10000$ ... is divisible by a divisor "**DV**". It would facilitate the divisibility analysis if **N** is reduced to a smaller number $N_{congruent}$ which has the same modular property as that of **N** when divided by "**DV**", that is: $N_{congruent} \equiv$ N **modulo DV**.

A simple way to achieve this goal is to replace the powers of **10** in the above equation of **N** by the remainders R_1, R_2, R_3, R_4 to R_{k-1} of the divisions

$$\frac{10}{DV}, \frac{10^2}{DV}, \frac{10^3}{DV}, \frac{10^4}{DV}, \frac{10^5}{DV} \text{ .. to } \frac{10^{k-1}}{DV}$$

respectively, where **k** is the number of digits in the numeral "**N**".

Accordingly, $N_{congruent} = D_0 + D_1* R_1 + D_2* R_2 + D_3* R_3 + D_4* R_4$ ...

The process of validating a divisibility testing method will proceed as follows:

a. The divisibility testing method is first presented and the testing procedure is initiated.

    A divisibility testing number DT_{DV} which is congruent with **N** modulo **DV** ($DT_{DV} \equiv$ N mod **DV**) is generated as a result of applying the testing procedure on the digits composing the numeral **N**. In simple words ($DT_{DV} \equiv$ N mod **DV**) means that if DT_{DV} is divisible by **DV**; **N** will also be (remember that **DV** is the divisor)

b. Validation is accomplished if we can prove that the divisibility testing number DT_{DV} is identical to or congruent with $N_{congruent}$ modulo **DV** ($DT_{DV} = N_{congruent}$ OR $DT_{DV} \equiv N_{congruent}$ mod **DV**).

Some modularity restoring adjustments on DT_{DV} may be required for such goal to be accomplished. Adjustments may include adding or subtracting multiples of the divisor **DV** to/from the the divisibility testing number DT_{DV} or dividing it by any number which is co-prime with **DV** (shares no factors of it)

If it is proven that $DT_{DV} \equiv N_{congruent}$ mod **DV** as such; it will accordingly mean that $DT_{DV} \equiv N$ mod **DV** or in simple words: if DT_{DV} is divisible by **DV** then **N** will also be, hence the validity of the proposed divisibility testing method is confirmed.

I shall follow these steps here in the "Divisibility Test Tabulations" designed to validate the algorithms of examining divisibility by **DV** for **DV** = **3**, **7**, **9**, **11** and **13** but shall not strictly follow an ascending order for **DV**

A. DIVISIBILITY BY 2, 4, 5 AND 8

Testing the divisibility by **2, 4, 5** or **8** is relatively simple and probably self-validated.

- **Divisibility by 2**: If and only if the unit's digit in a number **N** is an even number (and 0 is an even number too) then **N** is divisible by **2**

- **Divisibility by 4**: If and only if the two-digit number formed by the units' and the tens' digits (D_1 & D_O at rightmost part of **N**) is divisible by **4** then **N** will also be.
  For example: **25012** is divisible by **4** because **12** is divisible by **4**

- **Divisibility by 5:** If – and only if - the unit's digit in a number **N** is **0** or **5** then **N** is divisible by **5**

- **Divisibility by 8**: If and only if the three-digit number formed by the units', the tens' and the hundreds' digits (D_2, D_1 & D_O at rightmost part of **N**) is divisible by **8** then **N** will also be.
  For example: **25312** is divisible by **8** because **312** is divisible by **8**
  A three digit number – formed by the digits D_2, D_1 & D_O - will be divisible by **8** in the following two cases:

  i. If D_2 is odd and the two-digit number formed by D_1 & D_O (at the rightmost part of **N**) is divisible by **4** and **NOT** divisible by **8**.
     Example: **312**; where 3 is odd and **12** is divisible by **4** (and **NOT** divisible by **8**)

ii. If D_2 is even and the two-digit number formed by D_1 & D_0 (at the rightmost part of **N**) is divisible by **8**.

Example: **216**; where **2** is even and **16** is divisible by **8**

B. DIVISIBILITY BY 9 AND 3

Let us examine the divisibility of the number given in the above example (**24,876**) by **9**.

The rule of divisibility by 9 is that a number N is divisible by 9 if and only if the sum of digits composing N is divisible by 9.

Sum of the digits in the example number is (**2 + 4 + 8 + 7 + 6 = 27**) which is divisible by **9**; hence number **24,876** is also divisible by **9**.

Such a special property of "**9**" is also shared by "**3**":

If and only if the sum of the digits of a number is divisible by 3 then the number itself is divisible by 3

What makes **9** and **3** privileged as such?

This is attributable to the fact that the base of decimal system;

$10 \equiv 1 \bmod 9 \equiv 1 \bmod 3$

(In simple words, **1** is the remainder of the divisions **10/9** and **10/3**)

Not only that the remainder of the division **10/ 9** is **1**, but the remainder of the division (**10^k / 9**) in general is **1**, where **k** is any positive integer (**100/ 9 = 11** with 1 as a remainder, **1000/ 9 = 111** with **1** as a remainder, **10000/ 9 = 1111** with **1** as a remainder, **100000/ 9 = 11111** with **1** as a remainder .. etc.)

Such modular property – in which the **remainder** of the division

10^k / 9 equals **1** is demonstrated in the tabulation in figure 5-2.

The property is also shared by **3** as a divisor

This modular property of **9** and **3** makes the sum of the digits in a number **N** represents the remainder of the division **N ÷ 9** (or **3**)

DIVISIBILITY TEST				DIVISOR "DV" =			9	
REMAINDER OF THE DIVISION $(10^i/9) = R_i$	$R_6 = 1$	$R_5 = 1$	$R_4 = 1$	$R_3 = 1$	$R_2 = 1$	$R_1 = 1$	$R_0 = 1$	
Ncongruent $\equiv$ N mod 9	$N_{congruent} = \sum_{0\,to\,i} (R_i * D_i) \equiv$ N mod 9 $= R_6{*}D_6 + R_5{*}D_5 + R_4{*}D_4 + R_3{*}D_3 + R_2{*}D_2 + R_1{*}D_1 + R_0{*}D_0$							
Ncongruent	$1{*}D_6 + 1{*}D_5 + 1{*}D_4 + 1{*}D_3 + 1{*}D_2 + 1{*}D_1 + 1{*}D_0$							
SUGGESTED DIVISIBILITY "TEST PROCEDURE" .	Method: Sum up the digits of the number N, $(D_0 + D_1 + D_2 + D_3 + D_4 + D_5 + ...)$ If the sum of digits is divisible by 9; N will also be .							
RESULTING DT_9	$DT_9 = D_6 + D_5 + D_4 + D_3 + D_2 + D_1 + D_0$							
CONCLUSION: DT_9 IS CONGRUENT TO $N_{congruent}$ HENCE CONGRUENT TO "N" SO THE METHOD IS VALID								

COMPARE THESE TWO EXPRESSIONS

Figure 5-2

C.   DIVISIBILITY BY 7

While testing the divisibility by **9** or **3** is a straight forward exercise, testing the divisibility by **7** requires little more work. Here are the steps of testing the divisibility by **7**:

a. Cut off the unit's digit – the first to right – from the number to be tested (N)

b. Subtract twice that unit digit from what remained in N after the removal of its unit's digit. Let the result of subtraction be DT_7

c. If the resulting number (DT_7) is divisible by seven; N will also be

d. If (DT_7) is still too big for you to examine its divisibility readily and without calculator, repeat steps a through c iteratively until you are satisfied with its modular status (with 7 being the modulus)

Let us examine - as an example – the divisibility of the number **539** by **7**. As discussed above, we would cut off the units' digit (**9**), double it and subtract this double from what has remained from original number: **53 – 2* 9 = 35**, and since **35** is divisible by **7** we conclude that the number **539** is also divisible by **7**.

Examining the divisibility of numbers having five or more digits will require an iterative process involving two or three steps like that.

Let us now examine **408793** for example; and I suppose it will not require much textual explanation:

To test **408793** : **40879 – 2* 3 = 40873**

then**40873** : **4087 - 2* 3** **= 4081**

then**4081** : **408 - 2* 1** **= 406**

then**406** : **40 - 2* 6 = 28;**

and since **28** is divisible by **7** we can conclude that **408793** is divisible by **7** as well.

The validation of this testing rule is outlined in figure 5-4; but I want to remind the reader that divisibility of **N** by a divisor **DV** is linked to the divisibility of **10, 100, 1000, 10000 ….** by **DV**

To examine the divisibility of **N** by **7**, we should first identify the modular properties of 10, 100, 1000, etc. with **7** being the modulus, that is the remainders of dividing **10, 100, 100, 1000** .. by 7 which is shown in the tabulation of figure 5-3

DIGITAL VALUE "DV"	1,000,000	100,000	10,000	1000	100	10	1
REMAINDER OF THE DIVISION (DV / 7)	1	5	4	6	2	3	1
	Recurring cycle		**Recurring cycle**				

Figure 5-3

Summing the products of the remainders in figure 5-3 and respective digits in a tested number will result in a smaller number (than N) which is modularly congruent to it modulo 7:

$N_{congruent} = D_0 + D_1{*}R_1 + D_2{*}R_2 + D_3{*}R_3 + D_4{*}R_4 + .. + D_{k-1}{*}R_{k-1}.$

This summation is also shown in the tabulation in figure 5.4

	DIVISIBILITY TEST DIVISOR "DV" = 7								
$N_{congruent}$	REMAINDER OF THE DIVISION $(10^i/7) = R_i$	$R_6 = 1$	$R_5 = 5$	$R_4 = 4$	$R_3 = 6$	$R_2 = 2$	$R_1 = 3$	$R_o = 1$	
			RECURRING CYCLE						
	Ncongruent ≡ N mod 7	$N_{congruent} = \sum_{0 \text{ to } i} (R_i * D_i) \equiv N \bmod 7$ $= R_6{*}D_6 + R_5{*}D_5 + R_4{*}D_4 + R_3{*}D_3 + R_2{*}D_2 + R_1{*}D_1 + R_O{*}D_O$							
DIVSIBILITY TEST	Ncongruent	$1{*}D_6 + 5{*}D_5 + 4{*}D_4 + 6{*}D_3 + 2{*}D_2 + 3{*}D_1 + 1{*}D_O$							COMPARE THESE TWO EXPRESSIONS
	SUGGESTED DIVISIBILITY "TEST PROCEDURE" .	(1) Cut off the units digit D_0 of the tested number N. (2) Subtract double of the cut off digit ($2 D_0$) from what remained of the number - If the resulting number (DT_7) is divisible by 7, N will also be							
TESTING PROCEDURE	RESULTING NUMBER: DT_7	$DT_7 = 100{,}000\ D_6 + 10{,}000\ D_5 + 1000\ D_4 + 100\ D_3 + 10\ D_2 + D_1 - 2\ D_O$							
	ADJUST DT_7 TO MAKE IT CONGRUENT TO $N_{congruent}$ & N	SUBTRACT 99995 D_6	SUBTRAC T 9975 D_5	SUBTRACT 980 D_4	SUBTRACT 70 D_3	NOTHING	ADD 14 D_1	ADD 7 D_O	
		$5\ D_6 +$	$25\ D_5 +$	$20\ D_4 +$	30 D3 +	$10\ D_2 +$	$15\ D_1 +$	$5\ D_0$	
		DIVIDE ALL THESE TERMS BY 5							
	ADJUSTED DT_7	$1{*}\ D_6+$	$5{*}\ D_5+$	$4{*}\ D_4+$	$6{*}\ D_3+$	$2{*}\ D_2+$	$3{*}\ D_1+$	$1{*}\ D_O$	
			RECURRING CYCLE						
CONCLUSION: DT_7 IS CONGRUENT TO $N_{congruent}$ HENCE CONGRUENT TO "N"									

Figure 5-4: Divisibility by 7- Testing Method and its validation

You will notice in the tabulation of figure 5-3 that the remainders of the divisions $\mathbf{10^{k-1}}$ **/ 7** are being repeated in a recurring cycle grouping six terms in each, so we will not need to include more than the 7 terms in the validation analysis shown in the tabulation of figure 5-4.

D. DIVISIBILITY BY 11

Another number that has special wonderful divisibility properties is 11, and I shall start by an example:

Consider the divisibility of **6893744** by **11**.

What we should do is to sum the digits at 1st, 3rd, 5th and 7th places and call the sum S_{ODD} then to sum the digits at 2nd, 4th and 6th places and call the sum S_{EVEN} (this refers to the sum of even ordered digits and has nothing to do with the number 7)

We would next find the arithmetic difference "R" between S_{ODD} and S_{EVEN}.

If and only if R = 0, 11 or a multiple of 11 then the tested number is divisible by 11.

$$\text{-- + -- + --------} = S_{EVEN} = 15$$
$$6\ 8\ 9\ 3\ 7\ 4\ 4$$
$$\text{-- + -- + -- + --} = S_{ODD} = 26$$

$$S_{ODD} - S_{EVEN} = 26 - 15 = 11$$

The divisibility analysis for **6893744** (to the right side) reveals that the number is divisible by **11**

What makes **11** so special in the properties of divisibility as such?

This is attributable to the fact that the base of decimal system; *10 ≡ -1 mod 11*

This means – and it may sound somewhat odd – that the remainder of the division **10/11** is **-1**

Continuing further in dividing higher powers of 10:

100 ≡ 1 mod 11 meaning that the remainder of the division **100/11** is **1** (or 100=9*11 + **1**),

1000 ≡ -1 mod 11 meaning that the remainder of the division **1000/11 is -1 (or 1000=91*11 - 1),**

10000 ≡ 1 mod 11 meaning that the remainder of the division **10000/11** is **1** (or 10000 = 109*11 + 1),

Because the remainder of the division **10^{r} /11** keeps alternating between **1** and **-1** as such; sum of evenly ordered digits in a number divisible by 11 is modularly congruent to the sum of oddly ordered digits $S_{ODD} \equiv S_{EVEN}$ (for S_{ODD} and S_{EVEN} to be modularly congruent - in this context - means that the difference between them is either 0, 11 or a multiple of 11)

Divisibility analysis and validation for 11 as a divisor is summarized in figure 5-5

DIVISIBILITY TEST				DIVISOR "DV" = 11			
REMAINDER OF THE DIVISION $(10^i / 11) = R_i$	$R_6 = 1$	$R_5 = -1$	$R_4 = 1$	$R_3 = -1$	$R_2 = 1$	$R_1 = -1$	$R_0 = 1$
$N_{congruent} \equiv N$ mod 11	$N_{congruent} = \sum_{0 \text{ to } i} (R_i * D_i) = R_6*D_6 + R_5*D_5 + R_4*D_4 + R_3*D_3 + R_2*D_2 + R_1*D_1 + R_0*D_0$ = $1*D_6 - 1*D_5 + 1*D_4 - 1*D_3 + 1*D_2 - 1*D_1 + 1*D_0$						
SUGGESTED DIVISIBILITY "TEST PROCEDURE"	Sparately add even ordered digits Deven= (D0 + D2 + D4 + D6 ..) and odd ordered digits Dodd= (D1 + D3 + D5 + D7 ..). If the difference between Deven and Dodd = 0 or any number divisible by 11; then N is divisible by 11						
RESULTING DT_{11}	$DT_{11} = (D_6 + D_4 + D_2 + D_0) - (D_5 + D_3 + D_1)$ $= D_6 - D_5 + D_4 - D_3 + D_2 - D_1 + D_0$						
CONCLUSION: DT_{11} IS CONGRUENT TO $N_{congruent}$ HENCE CONGRUENT TO "N"							

COMPARE THESE TWO EXPRESSIONS

Figure 5-5: Divisibility by 11 - Testing Method and its validation

E. DIVISIBILITY BY 13

The steps of examining the divisibility of N by 13 are somehow similar to those followed in examining the divisibility by 7. The process proceeds as follows:

a. Cut off the unit's digit – the first to right – from the number to be tested (N)

b. Add four times that unit digit to what remained in N after the removal of its unit's digit. Let the result of addition be **DT_{13}**

c. If the resulting number (**DT_{13}**) is divisible by **13**; **N** will also be

d. If (**DT_{13}**) is still too big for you to examine its divisibility readily and without calculator, repeat steps a through c iteratively until you are satisfied with its modular status (with **13** being the modulus)

Let us examine - as an example – the divisibility of the number **481** by **13**.

As discussed above, we would cut off the units' digit (**1**), quadruple it and add this quadruple to what has remained from original number: $DT_{13} = 48 + 4* 1 = 52$, and since **52** is divisible by **13** we conclude that the number **539** is also divisible by **13**.

In case DT_{13} would still be too large for its divisibility by 13 to be examined mentally, we would repeat the reduction process until we get a rather manageable value for it.

As was the case in testing divisibility by 7, the remainders of the divisions $10^{k-1} /13$ are being repeated in a recurring cycle grouping six terms in each, so we will not need to include more than the 7 terms in the validation analysis shown in the tabulation of figure 5-6 which summarizes the divisibility analysis and validation for 13 as a divisor.

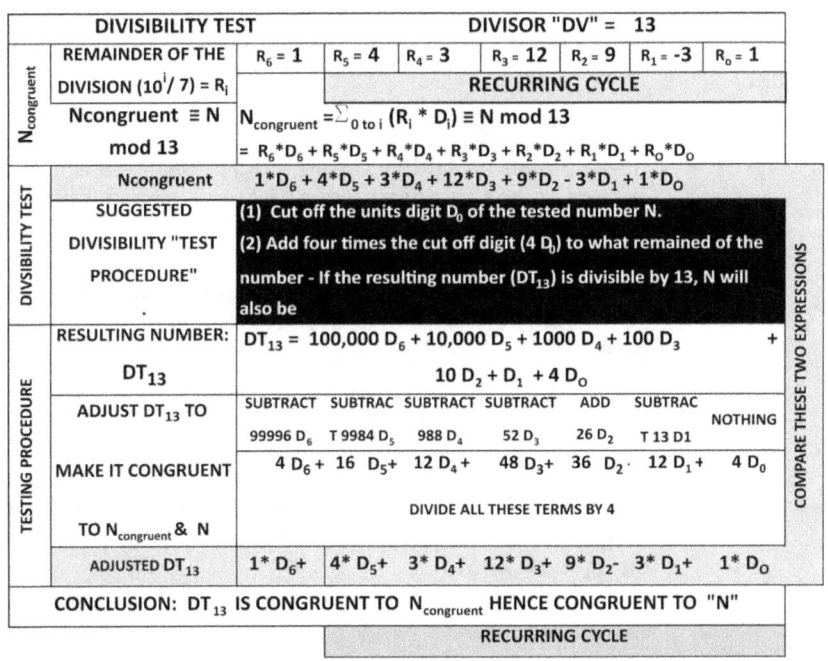

Figure 5-6: Divisibility by 13 - Testing Method and its validation

SIMULTANEUS DIVISIBILITY BY MORE THAN A DIVISOR

There is one more important general divisibility rule:

If **N** is divisible by two or more numbers then it should also be divisible by the "**Least Common Multiple**" of these numbers, and if these divisors are co-prime (they share no factors) then **N** should be divisible by the product of these numbers.

$$+ \quad + \quad = S_{EVEN} = 9$$

$$1\ \boxed{5}\ 3\ \boxed{4}\ 5$$

$$+ \quad + \quad = S_{ODD} = 9$$

With this rule in mind, as based on divisibility test methods discussed earlier in this chapter, we can extend our divisibility testing skill to include other composite divisors such as **6 = 2*3, 12= 3*4, 14 = 2*7, 18 = 2*9, 24 = 3*8, 26 = 2*13, 91= 7*13 and 99=9*11**

For example; take N = **15345** and see the analysis chart to the right side.

- Since the sum of oddly ordered digits S_{ODD} = **9** and the sum of evenly ordered digits S_{EVEN}=**9** also, then **N** is divisible by **11**

- And since the sum of all digits S_{ODD} + S_{EVEN} = **9** + **9** = **18** which is divisible by **9** then **N** must also be divisible by **9**

- And since the units digit is **5** then the **N** is divisible by **5**

- **11**, **9** and **5** are co-prime hence the Least Common Multiple= **11*9*5**= **495**, so it i concluded that **15345** is divisible by **495**

DO YOU KNOW IN WHICH DAY OF THE WEEK YOU WERE BORN?

One of modular arithmetic questions we often seek answers for is how to find the **day of the week** (that is Monday, Tuesday, Wednesday... etc.) for events such as anniversaries or birthdays and we often use tools such as a calendar or a cellular phone to assist us in finding the answers.

I am going to introduce here a simple method to find the **day of the week** for any date in the **20th** or the **21st** century (from **1st January 1901** to **31st December 2099**)

According to this method; you can find the **Week Day Number** (let us call it **WDn**) by getting the **modular sum** of the **Year Number Yn**, the **Month Number Mn** and the **Day Number Dn**, where **Yn**, **Mn** and **Dn** are key numbers determined by the year, the month and the day in which the event takes place. **Yn**, **Mn** and **Dn** are tabulated in context of discussion and are easy to memorize but I must explain how they have been calculated in the first place.

This modular sum that identifies the week day is given the name Weekday number (**WDn**). It ranges between **0** and **6** for days from **Sunday** through **Saturday** respectively.

Let us take an example to guide us through the process of identifying weekdays. Suppose – for example - that you want to know the **day of the week** at the breakout of world war - II on **1st September 1939**.

a. YEAR NUMBER

If you celebrate your birthday in a certain year on **Monday** (for example), you will celebrate it after **365** days – **which is obviously a simple year** - on **Tuesday**, because **365 = (7 * 52) + 1** hence there will always be one-day difference between two events separated by a **simple year**. If that year following the former celebration will **include the 29th of February**, it will be a leap year having a length of **366** days upon the elapse of which your birthday will be on **Wednesday**. That is because **366 = (7 * 52) + 2**. So, there will be **two days** difference between two events separated by a **leap year**
This is the principle upon which the year number **Yn** is determined.

To calculate the Year Number consider only the **First Two Digits** (**FTD**) of the year taken from right hand side that is the number formed by the **units'** and **tens'** digits. In the example in question we should consider **39** (and not **1939**).

The Year Number **Yn** will be calculated by adding the **First Two Digits** (**FTD**) to the integral part of (**FTD/ 4**) with the fraction dropped. This is expressed mathematically as follows:

Yn = FTD + Int (FTD /4).

The sign "**Int**" stands for "*the integral part of the number*". For example, **Int (26 ÷ 4) = Int (6.50) = 6**

For an event that happened in the year **1997** (we would proceed with calculation in the same way as if it happened in **2097**, because we are only concerned with the rightmost two digits: **97**)

Yn = 97 + Int(97/ 4) = 97 + Int(24.25) = 97 + 24 = 121

(that is **97** plus the whole number of the division **97 / 4**)

We want to know the day of the week so our final result should only range from 0 to six. I would therefore refer you to our discussion about *modular addition* and remind the reader that: **121 = 2 mod 7**, which means that **Yn = 121** is congruent to **Yn = 2** with **7** being the modulus; because **121 = (7* 17 +2)**. This is expressed as: **121 ≡ 2 mod 7**

Calculating the **Year Number** or any other variable in this process can be made simpler by dropping the multiples of **7** from any interim or final result.

Yet; there is a minor but important adjustment to the year number calculation in case the event comes in January or February of a leap year in which Yn is divisible by 4. In such a case we should deduct 1 from Yn.

For example, the year number **Yn** for an event that occurs in **January or February 2016** is: **16 + (16/4) - 1 = 19** which is congruent to **19 - 2*7 = 5**

Back to our example in which the weekday of **1ˢᵗ September 1939** was sought:

In that example the Year Number **Yn** = FTD + **Int (FTD /4)** = **39 + Int (39/4)** = **39 + 9 = 48** ≡ (**48** − **6*7**) = **6**. (in this last step; multiples of **7** were dropped)

Hence **Yn = 48 ≡ 6**

THE MONTH NUMBER Mn	
MONTH	MONTH NUMBER Mn
January October	0
February March November	3
April July	6
May	1
June	4
August	2
September December	5

Figure 5-7

b. MONTH NUMBER

Assume that the month number for the month of **January** is given the value **Mn = 0**. The month number for the month of **February** must obviously be equal **0 + 31 = (7 * 4) + 3 ≡ 3** because there are 31 days in January.

Similarly, the month number for the month of **March** in a simple year must be equal to **3 + 28 = (7 * 4) + 3 ≡ 3** also because there are 28 days in February that comes in a simple year (remember that the Year Number already accounts for the inclusion of leap years as explained).

Similarly, the month number for the month of **April** must be equal to **3 + 31 = (7 * 4) + 6 ≡ 6** because there are 31 days in March. **Mn** for the rest of the months can be worked out similarly and are listed in the tabulation of figure 5-7.

From the tabulation of figure 5-7; the month number **Mn** that corresponds to the month of **September** is **5**

c. DAY NUMBER

The day number **Dn**, is simply the calendar day of the month in which the event occurs or its modular equivalent.
In our example about the out the outbreak of World War II; **Dn = 1** for the event occurring on the first day of the month.

d. WEEKDAY NUMBER

Weekday number **WDn** is the final outcome of our calculations which ranges between **0** and **6**.
It is the **modular sum** of the Year Number **Yn**, Month Number **Mn** and Day Number **Dn**.

Remember that the **modular sum** means the arithmetic sum (**Yn +Ym + Yd**) with multiples of seven subtracted therefrom.

Weekday number ≡ (**Year Number + Month Number + Day Number)**

WDn ≡ (Yn + Mn + Dn)
IMPORTANT NOTE:
There is an important adjustment to the Weekday Number calculation:
In case the date calculated is in the 21st century (20XX), subtract 1 from WDn

In the given example: **WDn = (Yn + Mn + Dn) = 6 + 5 + 1 ≡ 12 ≡ 5**

Last step in our search is to find the weekday that corresponds to **WDn = 5** in the table of figure 5-8

WEEKDAY NUMBER WD$_n$	
DAY	WD$_n$
Monday	1
Tuseday	2
Wednesday	3
Thursday	4
Friday	5
Saturday	6
Sunday	0

Figure 5-8

From the tabulation in figure 5-8, **WDn = 5** corresponds to **Friday**.

Accordingly, **1st September 1939**, the day marking the breakout of World War II was **Friday**.

Note:
Had we been searching the weekday for **1st September 2039**, we should have subtracted **1** from **WDn** so it becomes **5 − 1 = 4;** as directed in the above **IMPORTANT NOTE**.
WDn = 4 corresponds to **Thursday** (according to the tabulation in figure 5-8), hence **1st September 2039** will be **Thursday**

For your convenience; if you do not like calculations at all – though you are reading a book on mathematics – I have developed for you and present in figure 5-9, a fantastic *200 years calendar*

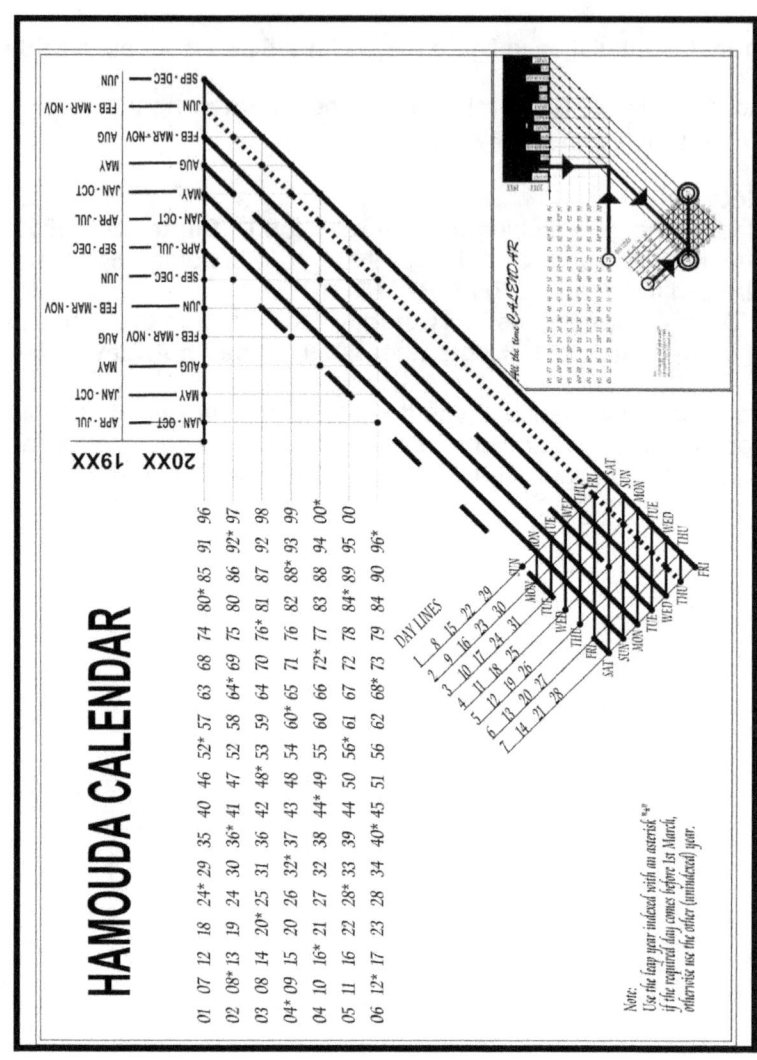

Figure 5-9 :

Hamouda Calendar - For any date in the 20th or 21st centuries

The Chapter Quiz

Prove that $(a^7 - a)$ is divisible by 7 for a = any positive integer (whole number)

Discussion and solution of the Chapter Quiz

This problem is in fact a special case of the general theorem known as **Fermat's Little Theorem** stating that:

$(a^p - a)$ is divisible by p for "**p**" being any prime number and "**a**" being any integer which is co-prime with "**p**".

The theorem and its proof are addressed in the detailed discussion of Chapter Seven's quiz but the proof of this special case in question takes a different path involving less intricate analysis.

The technique typically used to solve this kind of problems is the **Mathematical Induction**.

Steps taking place in applying this technique involve the following:

I. Show that the formula to be proved is valid for **a =1**
II. Prove that if divisibility is valid for **a = r** ; it will also be valid for **a = r + 1**

Performing the first step means that the formula is valid for **a = 1**, and performing the second step means that since it is valid for a = 1 it should also be valid for **a = 1 + 1 = 2**, henceforth it is valid for **2 + 1 = 3** and so on. This confirms validity for **a** = any positive natural number.

Let us start and assume that f(**a**) = $(a^7 - a)$

First step: Substitute for a=1 in the equation f(**a**) = $(a^7 - a)$: f(**a**)

$= 1^7 - 1 = 0$

0 is divisible by all numbers since **a * 0 = 0** hence **0/ a = 0**.

However if you do not feel comfortable about the divisibility of 0, move forward to a = 2 in which case **f(2) = (2^7 – 2) = 126** which is divisible by 7, so let us move forward to the second step.

We assume that the expression is valid for **a = r**, so **f(r) = (r^7 – r)** is divisible by **7**

Now let us examine the divisibility of **f(r + 1) = {(r + 1)^7 – (r + 1)}**

The binomial expansion of the right hand side of the above equation is:

$$(r + 1)^7 - (r + 1) = \{(1 + 7r + \frac{7*6}{2*1}r^2 + \frac{7*6*5}{3*2*1}r^3 + \frac{7*6*5*4}{4*3*2*1}r^4$$

$$+ \frac{7*6*5*4*3}{5*4*3*2*1}r^5 + \frac{7*6*5*4*3*2}{6*5*4*3*2*1}r^6 + r^7) - (r + 1)\}$$

hence **f(r+1)** $= 7 (r + \frac{6}{2*1}r^2 + \frac{6*5}{3*2}r^3 + \frac{6*5}{3*2}r^4 + \frac{6*5}{5*2}r^5 + r^6)$

+ (r^7 – r)

7 is multiplied by the expression in the first bracket, while the second bracket **(r^7 – r)** has initially been assumed to be divisible by **7** hence **f(r + 1)** is divisible by **7.**

Step **II** in the **Mathematical Induction** process has thus been successfully accomplished and **(a^7 – a)** is proven to be divisible by **7** for all integral values of "**a**"

Chapter 6: π

Mathematicians realized that the circumference to diameter ratio in a circle is a fixed constant since no less than 4000 years. Exhaustive efforts have been exerted ever since to find out what this ratio is – numerically. It was not until the seventeenth century that it was established that pi - often written "π" which is a Greek letter pronounced as "p" - was an irrational number whose exact value did not exist and could not be found.

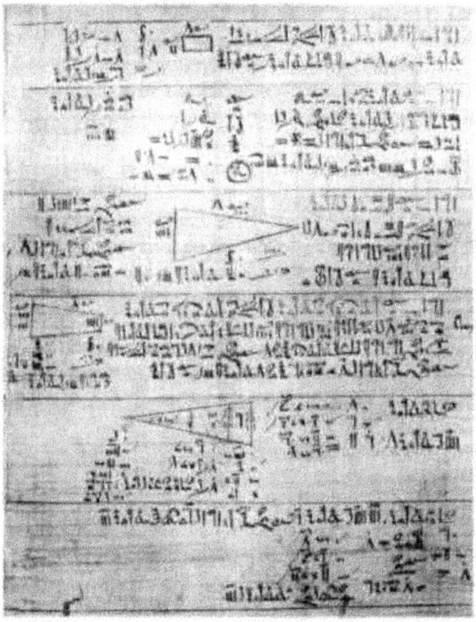

The first known mathematician who had input in the determination of an approximate value of π is Ahmes, whose scribe was written in an Egyptian papyrus around 1650 BC and was – according to Ahmes himself- copied from another scribe that was dated +200 years before Ahmes' scribe (figure 6-1)

Figure 6-1: Ahmes' scribe

Ancient Egyptian surveyors developed the geometry and used it as a tool to estimate the area of land inundated by the flood every year; mainly to assess the tax or compensation to be applied with land owners.

Ahmes' papyrus is also known as the Rhind papyrus after Alexander Rhind the 19th century antiquarian who purchased it in 1856 from Luxor, Egypt. Now it is exhibited in the British museum

The Rhind papyrus lists 87 mathematical problems, but the one that concerns us in this context is problem 50 which reads as follows:

"To calculate the area of a circular field of 9 arms diameter, take away 1/9 the diameter then square the balance 8; to find that its area= 64"

According to Ahmes; the area of a circle

$$A = \left(\frac{8}{9}D\right)^2 = 0.79\,D^2 = 3.16\,r^2$$

where **r** is the radius of the circle = ½ **D**.

This means that the value of π as approximated by **Ahmes** is **3.16**, which is nearly **0.0184** in excess of the modern approximation of (**3.141593 ...)**

Almost at the same period of Ahmes, the ancient Babylonians had their own assessment for π.

The Babylonian's Collection of Yale University holds a clay tablet dated between 1800 and 1600 BC (figure 6-2) revealing that the Babylonians established that the circumference of a circle was three folds of its diameter. This means that they estimated π to be **3.0**

Figure 6-2:

A Babylonian tablet

The approximation of π by the ancient Egyptians and Babylonians was strictly empirical based on measurements and not analysis.

That had been the case for long time until the polygonal perimeter approximation algorithm was introduced by **Archimedes of Syracuse** (287 – 212 BC)

The algorithm involved the computation of the perimeters of circle inscribed **and** circumscribed regular polygons of various side numbers (figure 6-3).

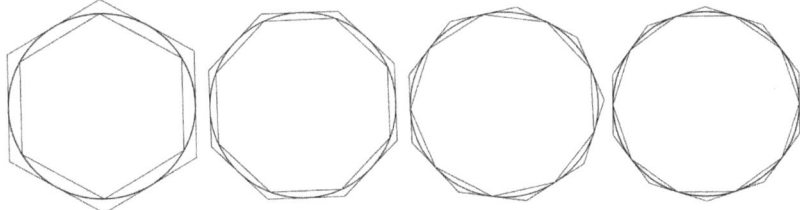

Figure 6-3

Archimedes' approximation is strictly based on geometric analysis; because neither the algebra nor the trigonometry had yet been developed at his time.

Figure 6-4: Archimedes

Even the decimal notation had not been in place then and he used the rational fractions (numerator/ denominator).

Archimedes' analysis started by calculating the perimeter of a hexagon inscribed in a given circle and that of another one circumscribed about it (figure 6-5). The circle circumference is obviously shorter than the perimeter of the circumscribed hexagon but longer than the perimeter of the inscribed one, and from here the approximation comes in the form of the given inequality.

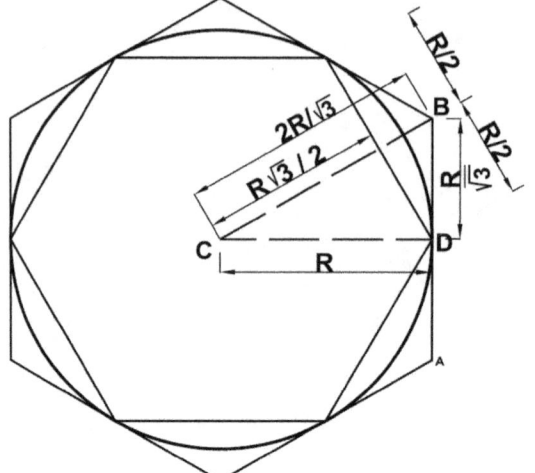

Figure 6-5:
Polygonal
perimeter
approximation

The perimeters of inscribed and circumscribed hexagons shown in figure 6-5 are: $6R$ and $4\sqrt{3}\,R$ (R being the radius of the circle) respectively.

Corresponding polygon perimeter/ circle diameter of inscribed and circumscribed hexagons are: **3.0** and **3.461** hence the approximation of π initially obtained so far is:

3.0 < π < 3.461

Archimedes then proceeded with the first iterative step to enhance the approximation; which is doubling the number of sides of the polygon from six to twelve. In absence of any trigonometric or algebraic techniques, Euclid's Angle Bisector proposition was quite useful.

This is Proposition 3 of book VI of the Elements – the wonderful encyclopedic reference of Euclid- states::

"If an angle of a triangle is bisected by a straight line cutting the base, then the segments of the base have the same ratio as the remaining sides of the triangle"

Angle **ACB** in triangle **ABC** (figure 6-6) is bisected by **CD**.

Euclid's Proposition says that:

$$\frac{CA}{CB} = \frac{AD}{BD}$$

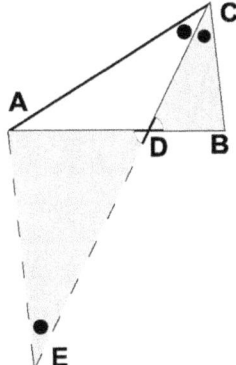

Figure 6-6

To prove the proposition; draw line segment **AE** parallel to **CB** and extend **CD** to meet **AE** at **E** as shown in the figure. Triangles **CBD** and **EAD** (shaded) are similar because ∠**AED** = ∠**DCB** (alternate) and ∠**CDB** = ∠**EDA** (opposite angles meeting at the intersection of two lines), hence:

$$\frac{EA}{CB} = \frac{AD}{BD}$$, and angle AED = angle BCD = angle ACD

It follows that triangle AEC is an isosceles triangle hence **CA = EA** . Replacing **EA** by **CA** in the above equation we get: $$\frac{CA}{CB} = \frac{AD}{BD}$$

The proposition is therefore proven so let us see how it was used by Archimedes to calculate the perimeter of the polygon after doubling the number of its sides.

The central angle facing any side of the regular hexagon is 60°. To double the number of sides, i.e. to convert the hexagon to 12 side regular polygon; central angle should be halved.

In figure 6-7; **AB** is one side of the hexagon whose center is "**C**".

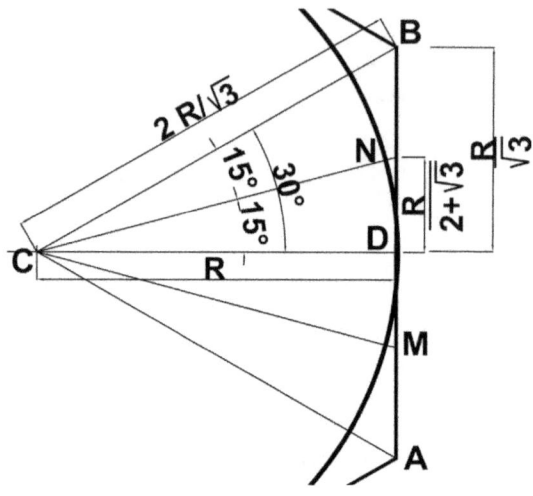

Figure 6-7

CN and **CM** are the bisectors of angles **BCD** and **ACD** respectively. **MN** is therefore a side in the new 12-gon that will be constructed and **DN** is a half of that side. To find the new 12 sided perimeter, we should find **DN**; which is attainable by using Euclid's proposition. Consider triangle **BCD**, and the bisector **CN** of the angle **BCD** and apply the proposition:

$$\frac{DN}{NB} = \frac{CD}{CB}$$

$$\frac{DN}{DN + NB} = \frac{CD}{CD + CB} \; ; \; \frac{DN}{DB} = \frac{CD}{CD + CB}$$

$$DN = DB * \frac{CD}{CD + CB}$$

$$DN = \frac{R}{\sqrt{3}} * \frac{R}{R + \frac{2R}{\sqrt{3}}} = \frac{R}{2 + \sqrt{3}}$$

DN is half of the side of the **12** sided polygon. The ratio: (perimeter of circumscribing polygon) / (circle diameter) will be equal to

(24 DN)/2R = $\frac{12}{2+\sqrt{3}}$ **= 3.21539**

This is better than the approximation resulted from the circumscribing hexagon; which was **3.461**

Before another cycle of doubling the number of the sides of circumscribed polygon can take place, **CN** should be obtained using the Pythagorean Theorem: **$CN^2 = CD^2 + DN^2$.**

Since the Algebra had not been developed at the time of Archimedes, his analysis was based on numerical values. In the example illustrated in figure 6-7; the radius of the circle might have been given the value **1.0** rather than **R**. Furthermore, approximated rational fractions must have been used. For example, might have been used in lieu of $\sqrt{3}$

Number of sides in the inscribed hexagon should also be doubled to **12** in the same manner followed to $\frac{97}{56}$ double the number of sides of circumscribing hexagon in above example.

After doubling the numbers of sides for both circumscribed and inscribed hexagons, the process was iteratively repeated by Archimedes to further doubling the number of sides from 12 to 24, then to 48 and finally to 96 sides. At this point, he came up with the following approximation of π :

$$\frac{223}{71} < \pi < \frac{22}{7}$$

The development of calculus in the seventeenth century by Isaac Newton and Leibniz led to the discovery of several infinite series that constituted a breakthrough in the quest of approximating π.

Isaac Newton (1642 – 1726) himself had a prominent role in the approximation efforts that was ongoing in the 17th and 18th centuries.

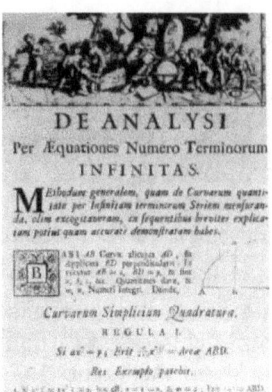

Figure 6-8: Isaac Newton

Figure 6-9: De Analysi

He wrote a paper called "**De analysi**" on the analysis by equations with infinitely many terms, and another one called: "A Treatise on the Methods of Series and Fluxions" (Katz 2009) which paved the road to a genius approximation algorithm.

In the De analysi, Newton explains how to calculate by - the so called fluxions - the area beneath the curve $y = ax^{m/n}$ above the **x** axis between points (**0** , **0**) and (**X** , **0**) - see figure 6-10 - and concludes that such area equals:

$$\frac{an}{m + n} x^{(m+n)/n}$$

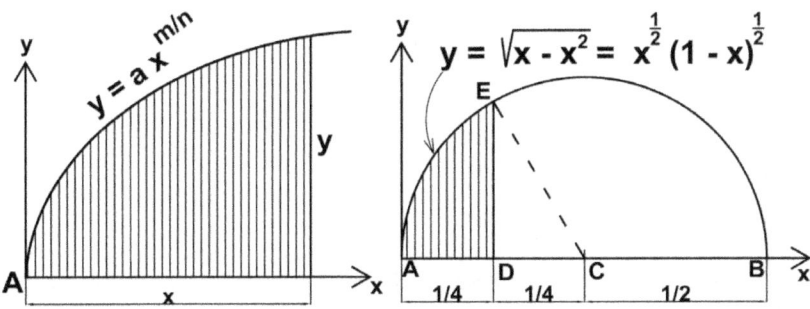

Figure 6-10   Figure 6-11

In modern mathematical notations; the solution obtained by fluxions is equivalent to the integration:

$$\int_0^X ax^{m/n} \, dx = a.\frac{x^{(m/n\ +1)}}{\frac{m}{n}+1}$$

Approximation analysis starts with drawing the semicircle **AEB** in which point **A** is the point of origin (**0** , **0**) and **B** is (**1** , **0**), hence the radius **EC= 1/2** as shown in figure 6-11.

The equation of the circle is

$$(x-\tfrac{1}{2})^2 + (y-0)^2 = (\tfrac{1}{2})^2,$$

$$\text{or } x^2 - x + \frac{1}{4} + y^2 = \frac{1}{4}$$

$$\text{hence } y^2 = x - x^2 \text{ or}$$

$$y = x^{\frac{1}{2}}.(1-x)^{\frac{1}{2}}$$

The binomial expansion of $(1 + x)^n$

$$= 1 + \frac{nx}{1!} + \frac{n(n-1)x^2}{2!} + \frac{n(n-1)(n-2)x^3}{3!}$$
$$+ \frac{n(n-1)(n-2)(n-3)x^4}{4!} + \cdots$$

Newton expanded the expression $(1 - x)^{1/2}$ in the equation of the circle in the same manner

$$y = x^{\frac{1}{2}} \cdot \left(1 - \frac{1}{2}x - \frac{1}{8}x^2 - \frac{1}{16}x^3 - \frac{5}{128}x^4 - \frac{7}{256}x^5 \cdots\right)$$

$$y = x^{1/2} - \frac{1}{2}x^{3/2} - \frac{1}{8}x^{5/2} - \frac{1}{16}x^{7/2} - \frac{5}{128}x^{9/2} - \frac{7}{256}x^{11/2} \cdots$$

Applying the principle of fluxions on this modified equation of the circle; Newton calculated the area A_{AED} (hatched area beneath the curve in figure 6-11)

$$A_{AED} = \frac{2}{3}x^{3/2} - \frac{1}{2} \cdot \frac{2}{5} \cdot x^{5/2} - \frac{1}{8} \cdot \frac{2}{7} \cdot x^{7/2} - \frac{1}{16} \cdot \frac{2}{9} \cdot x^{9/2} - \frac{5}{128} \cdot \frac{2}{11} \cdot x^{11/2} \cdots$$

Newton then substituted $x = \frac{1}{4}$ in the above equation of the area A_{AED}

It was pretty genius of Newton to choose the value of $\frac{1}{4}$ for x, because:

    a -   It was easy for him to calculate x raised to fractional powers such as 3/2, 5/2, 7,2 .. etc., because

$$\left(\frac{1}{4}\right)^{\frac{3}{2}} = \frac{1}{8}, \quad \left(\frac{1}{4}\right)^{\frac{5}{2}} = \frac{1}{32}, \quad \left(\frac{1}{4}\right)^{\frac{7}{2}} = \frac{1}{128}$$

b. By giving the measure of ¼ to AD, DC will also be l/4 because the radius **CE=1/2**, which makes **cosine(ACE) = DC/CE = 0.5**.

This means that angle **ACE** is **60°**, or π/3 in radians and the area of sector **ACE** is exactly **1/6** the area of full circle.

Using nine terms of the above equation to approximate the hatched area **AED** (figure 6-11), Newton found that A_{AED} = **0.0767731**

It was not difficult for Newton after that to find the area of triangle **EDC** geometrically, since:

$EC^2 = ED^2 + DC^2$ (Pythagorean Theorem) hence

$$ED = \sqrt{EC^2 - DC^2} = \sqrt{\left(\tfrac{1}{2}\right)^2 - \left(\tfrac{1}{4}\right)^2} = 0.43301270189$$

hence the area of triangle **ECD** = (1/4)*(0.4330127)/2 = **0.05412659**

Area of the sector **ACE** = Shaded area **AED** + Area of triangle **EDC**

= 0.0767731 + 0.05412659 = **0.130899695**

Since the area of the sector **ACE** is **1/6** the area of the full circle, the area of the circle = **6*0.130899695 = 0.785398**. which equals π * $(CE)^2 = π * 0.5^2$

The approximated value of π estimated as such by using nine terms of the binomially expanded area equation (A_{AED}) is

π = **3.141592668**, which is correct to seven decimals

In 1671, the Scottish mathematician **James Gregory** (1638 – 1675) discovered the arctangent infinite series:

$$\tan^{-1}(x) = X - \frac{x^3}{3} + \frac{x^5}{5} - \frac{x^7}{7} + \frac{x^9}{9} + ..$$

This important series merits elaboration and a quick review of some trigonometric basics.

Consider the triangle shown in figure 6-12 in association with following trigonometric notions:

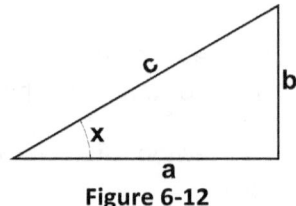

Figure 6-12

$$\sin(x) = \frac{opposite}{hypetenuse} = \frac{b}{c},$$

$$\textbf{\textit{hence arcsine}} \left(\frac{b}{c}\right) \textbf{\textit{[also called}} \sin^{-1}\left(\frac{b}{c}\right)] = x$$

$$\tan(x) = \frac{opposite}{adjacent} = \frac{b}{a},$$

$$\textbf{\textit{hence arctan}} \left(\frac{b}{a}\right) \textbf{\textit{[also called}} \tan^{-1}\left(\frac{b}{a}\right)] = x$$

So if **tan(x) = b/a** then x = tan⁻¹ (b/a)

Note:

tan⁻¹**(x)** and **arctan(x)** have the same meaning: "*the angle whose tangent, i.e. (opposite/ adjacent) equals x*"

We shall use **Taylor's expansion** technique to find the expansion of the inverse tangent function **tan**⁻¹**(x)** as an example that can also be followed with several other functions.

First step is to write the function in the general infinite polynomial form:

$$\tan^{-1}(x) = a_0 + a_1x + a_2x^2 + a_3x^3 + a_4x^4 + a_5x^5 + a_6x^6 + a_7x^7 + a_8x^8 \;....\; \infty,$$

Let us call it the "**base equation**"

We would then substitute **x = 0** in both sides of the base equation :

$\tan^{-1}(0) = a_0$ which means that $a_0 = 0$

Next step is to differentiate both sides of the base equation.

The first derivative (differentiation) of $\tan^{-1}(x)$ and X^n are :

$$\frac{d}{dx}\tan^{-1}(x) = \frac{1}{1+x^2} \text{ and}$$

$$\frac{d}{dx}\,x^n = n\,x^{n-1} \qquad \text{respectively}$$

Accordingly, the differentiation of the two sides of the base equation will result in creating the following equation:

$$\frac{1}{1+x^2} = a_1 + 2a_2\,x + 3\,a_3\,x^2 + 4\,a_4\,x^3 + 5\,a_5$$
$$x^4 + 6\,a_6\,x^5 + 7\,a_7\,x^6 + ...$$

Let us call it the **first derivative equation.**

Again by substituting x = 0 in both sides of the first derivative equation; we get: **1 = a_1**

We would further differentiate both sides of the first derivative equation and would get the following equation as a result:

$$-\frac{2x}{(1+x^2)^2} = 2a_2 + 6\,a_3\,x + 12\,a_4\,x^2 + 20\,a_5\,x^3 + 30\,a_6\,x^4 + 42\,a_7\,x^5$$
....

Substituting x=0 in both sides of the equation; we get: **0 = a$_2$**.

Proceeding with a third cycle of differentiation we get:

$$-\frac{2(1+x^2)^2 - 8x^2(1+x^2)}{(1+x^2)^4}$$

$$= 6\,a_3 + 24\,a_4\,x + 60\,a_5\,x^2 + 120\,a_6\,x^3 + 210\,a_7\,x^4 \quad ....$$

Substituting x=0 in both sides of the equation; we get: **2 = 6 a$_3$** hence **a$_3$ = 1/3**

With further **differentiation/ zero substitution** cycles we shall find:

a$_4$ = 0 , a$_5$ = 1/5 , a$_6$ = 0 , a$_7$ =1/7, etc. hence the pattern has shown up clearly, and the **Taylor's expansion** for $tan^{-1}(x)$ validates James Gregory's series:

$$\tan^{-1}(x) = x - \frac{x^3}{3} + \frac{x^5}{5} - \frac{x^7}{7} + \frac{x^9}{9} - \frac{x^{11}}{11} + \frac{x^{13}}{13} - \frac{x^{15}}{15} + \frac{x^{17}}{17}$$
.... (I)

Knowing that

$$\frac{d}{dx}\sin(x) = \cos(x) \text{ and } \frac{d}{dx}\cos(x) = -\sin(x);$$

we can follow the same procedure to find the polynomial expansion for the functions **sin(x)** and **cos(x)** which are written below.

We will need these expansions later in our discussions.

$$\sin(x) = x - \frac{x^3}{3!} + \frac{x^5}{5!} - \frac{x^7}{7!} + \frac{x^9}{9!} - \frac{x^{11}}{11!} + \frac{x^{13}}{13!} - \frac{x^{15}}{15!} + \frac{x^{17}}{17!} \ldots\ldots \quad \text{(II)}$$

$$\cos(x) = 1 - \frac{x^2}{2!} + \frac{x^4}{4!} - \frac{x^6}{6!} + \frac{x^8}{8!} - \frac{x^{10}}{10!} + \frac{x^{12}}{12!} - \frac{x^{14}}{14!} + \frac{x^{16}}{16!} \ldots\ldots \quad \text{(III)}$$

In the expansion of $\tan^{-1}(x)$ - equation (I) - if the angle $x = 45°$ in degrees which is $\pi/4$ in radians; $\tan(x)$ will be **1.0**; hence:

$$\pi/4 = 1 - \frac{1}{3} + \frac{1}{5} - \frac{1}{7} + \frac{1}{9} - \frac{1}{11} + \frac{1}{13} - \frac{1}{15} \ldots \quad \infty$$

The problem with this infinite series is that it converges very slowly. Summing the first 1000 terms will keep the approximation nearly 0.01 away from a more precise 14 decimal place approximation

As calculations were performed using primitive tools and only geometric algorithm, time and effort exerted in approximation with such a slow convergence were prohibitive. Speedy convergence was therefore of a significant importance. The slow convergence is attributable to fact that **1.0** raised to any power will remain **1.0**, hence terms of Gregory's series would diminish slowly for the substitution **x = 1**. This can be mitigated by splitting the $\pi/4$ angle into two smaller angles $\tan^{-1}(a)$ and $\tan^{-1}(b)$ such that $\pi/4 = \tan^{-1}(a) + \tan^{-1}(b)$.

Such splitting of $\pi/4$ is performed using the formula:

$$\tan(\tfrac{\pi}{4}) = \tan(a + b) = \frac{\tan(a) + \tan(b)}{1 - \tan(a).\tan(b)}$$

Following are a few examples of such splitting:

$\pi/4 = \tan^{-1}(1/2) + \tan^{-1}(1/3)$

$\pi/4 = \tan^{-1}(0.6) + \tan^{-1}(0.25)$

$\pi/4 = \tan^{-1}(0.8) + \tan^{-1}(1/9)$

The following converging splitting combination was developed in 1706 by the British astronomer **John Machin** (1686 – 1751).

It is much efficient and speedy converging.

$\pi/4$ = 4 tan^{-1}(1/5) - tan^{-1}(1/239)

The combination significantly enhances the convergence speed such that it is still used nowadays by computer programmers seeking additional new digits in the quest of approximating π.

Before proceeding any further with discussion I shall have to prove

Machin equation using the angle summation formula which was

referred to above.

The proof proceeds in the following three angle summation steps:

$$\tan[2\ \tan^{-1}(1/5)] = \tan[\tan^{-1}(1/5) + \tan^{-1}(1/5)] \quad = \frac{\frac{1}{5} + \frac{1}{5}}{1 - \frac{1}{5} * \frac{1}{5}} = \frac{5}{12}$$

$$\tan[4\ \tan^{-1}(1/5)] = \tan[2\ \tan^{-1}(1/5) + 2\ \tan^{-1}(1/5)] = \frac{\frac{5}{12} + \frac{5}{12}}{1 - \frac{5}{12} * \frac{5}{12}} = \frac{120}{119}$$

$$\tan[4\ \tan^{-1}(1/5) - \tan^{-1}(1/239)] \quad = \frac{\frac{120}{119} - \frac{1}{239}}{1 + \frac{120}{119} * \frac{1}{239}} = \frac{28561}{28561} = 1$$

The angle whose tangent = 1 is **45°** in degrees which is $\pi/4$ in

radians, hence $\pi/4$ = 4 tan^{-1}(1/5) - tan^{-1}(1/239)

It is amazing that a precision exceeds 14 decimal places has already been achieved in such an efficient algorithm by only summing up the first 100 terms of Gregory's series.

A remarkable breakthrough in expressing π as an infinite series was accomplished in 1748 by the great Swiss mathematician **Leonard Euler** (1707 – 1783) who was the first one to give the symbol π its name.

Leonhard Euler introduced the following amazing infinite series:

$$\frac{\pi^2}{6} = \frac{1}{1^2} + \frac{1}{2^2} + \frac{1}{3^2} + \frac{1}{4^2} + \frac{1}{5^2} + \frac{1}{6^2} + \cdots \quad \infty$$

$$= \sum_{n=0}^{\infty} \left(\frac{1}{n^2}\right)$$

The elegant proof brought by Euler merits a presentation here.

The algorithm followed by Euler in developing the formula is to expand a certain function f(x) in two different ways and to equate the resulting expansions.

Before proceeding further with the steps; I would like to pick an introductory example.

Figure 6-13: Leonhard Euler

If we know that the solutions of equation **g(x) = 0** are – for example - **x = 1**, **x = 3** and **x = 5** then we can form the relevant equation that yields these three solutions (also called the roots of "**x**") as follows:

g(x) = c (x - 1)(x - 3)(x - 5) = 0 , where **c** is an arbitrary constant.

Now let us apply this principle at the function

$$f(x) = \frac{\sin(x)}{x} = 0$$

The roots of this equation, i.e. the values of **x** that makes **f(x) = 0** are:

x = ± π, x = ± 2 π, x = ± 3 π, x = ± 4 π, x = ± 4 π ..etc .

You may notice that **x = 0** was excluded because the denominator **x** cannot be given a 0 value.

We can assume that:

$f(x) =$

$$\left(1 + \frac{x}{\pi}\right) \cdot \left(1 - \frac{x}{\pi}\right) \cdot \left(1 + \frac{x}{2\pi}\right) \cdot \left(1 - \frac{x}{2\pi}\right) \cdot \left(1 + \frac{x}{3\pi}\right) \cdot \left(1 - \frac{x}{3\pi}\right) \cdot \left(1 + \frac{x}{4\pi}\right) \cdot \left(1 - \frac{x}{4\pi}\right)$$
.....(A)

We have just expanded **f(x)** using the first method..

In the second method we will be using **Taylor**'s expansion of the function **sin(x)** (see equation II above)

$$\sin x = x - \frac{x^3}{3!} + \frac{x^5}{5!} - \frac{x^7}{7!} + \frac{x^9}{9!} - \frac{x^{11}}{11!} \ldots \quad \text{hence:}$$

$$f(x) = \frac{\sin x}{x} = 1 - \frac{x^2}{3!} + \frac{x^4}{5!} - \frac{x^6}{7!} + \frac{x^8}{9!} - \frac{x^{10}}{11!} \ldots \infty \quad \ldots\ldots\ldots \text{(B)}$$

Equating the right sides of equations **(A)** and **(B)** we get:

$$\left(1 - \frac{x^2}{\pi^2}\right) \cdot \left(1 - \frac{x^2}{4\pi^2}\right) \cdot \left(1 - \frac{x^2}{9\pi^2}\right) \cdot \left(1 - \frac{x^2}{16\pi^2}\right) \cdot \left(1 - \frac{x^2}{25\pi^2}\right) \ldots =$$

$$1 - \frac{x^2}{3!} + \frac{x^4}{5!} - \frac{x^6}{7!} + \frac{x^8}{9!} - \frac{x^{10}}{11!} + \frac{x^{12}}{13!} \ldots\ldots\ldots \text{(C)}$$

Now compare and equate the coefficients of X^2 in the two sides of equation (c).

Left side of equation (c) comprises the product of an infinite number

of brackets in the form:

$$\left(1 - \frac{x^2}{r^2\pi^2}\right)$$

In the expansion of the left side; terms containing X^2 are the product of the second term in **ONE** of the brackets and the first terms (**1**) in all others.

Equating the coefficients of in the two sides of equation (c) after expansion:

$$-\frac{1}{\pi^2} - \frac{1}{4\pi^2} - \frac{1}{9\pi^2} - \frac{1}{16\pi^2} - \frac{1}{25\pi^2} - \frac{1}{36\pi^2} \cdots = -\frac{1}{3!}$$

Hence,

$$\frac{-1}{\pi^2}\left(1 + \frac{1}{4} + \frac{1}{9} + \frac{1}{16} + \frac{1}{25} + \frac{1}{36} \cdots\right) = -\frac{1}{3!}$$

$$\frac{\pi^2}{6} = \frac{1}{1^2} + \frac{1}{2^2} + \frac{1}{3^2} + \frac{1}{4^2} + \frac{1}{5^2} + \frac{1}{6^2} + \frac{1}{7^2} + \cdots \infty$$

And this is nothing but the notorious infinite series of Euler

Euler series offers a slightly better convergence than that provided by Gregory's series in its simple application; but not when the conversion $\pi/4 = 4\tan^{-1}(1/5) - \tan^{-1}(1/239)$ is applied.

While we have just been discussing a great achievement of **Leonard Euler**, it would be appropriate to discuss another outstanding work of him which is the renowned **Euler's Identity** that reads:

$$e^{i\pi} + 1 = 0$$

Euler's Identity – also called Euler's Formula - is amazing, because it links π to **e** – which also happened to be called Euler's Number - and combines real and imaginary numbers.

The mysterious "e"

You have just seen that amid our discussion about "π", the mysterious mathematical constant "e" popped up unexpectedly, which made it inappropriate for me to ignore it and continue the discussion without giving a brief introductory note about it.

"e" – sometimes called Euler's number - is an important mathematical constant and irrational number that equals **2.71828...** (approximately).

Where did this number come from and what does it mean?

Imagine that you have deposited one dollar in a bank that pays an annual interest of "**i**" dollars. The deposited fund will amount to (**1 + i**) dollars after one year, **(1 + i)*(1 + i)** dollars after two years, **(1 + i)(1 + i)(1 + i)** after three years and – in general:

(1 + i)^n dollars after **n** years.

Suppose that the bank decided to raise the annual interest rate to 100% (yes it requires wild imagination to think of that), the deposited amount will be (**1 + 1**) dollar after one year.

Now imagine that you have reached an agreement with the bank upon which the same rate of annual interest (**100**%) will remain in effect but the interest will be calculated and accrued every month (instead of every year). Since the annual interest is **100%,** you should expect that monthly interest would be **100/ 12 = 8.3333**%, hence in **12** months, the $ **1.0** deposit will become:

$$\left(1 + \frac{1}{12}\right)^{12} = 2.61303$$

You will be encouraged by the fact that reducing the period of computation from one year to one month has caused the value of your investment to jump from $2.0 to $2.613 in one year, and would ask for further splitting of assessment period.
You will have a deal with the bank to keep the annual interest rate as agreed but to calculate and accrue the interest **every single day** (for another new deposit).

Accordingly, after one year of depositing; your balance account will become:

$$\left(1 + \frac{1}{365}\right)^{365} = 2.71456$$

You will be further thrilled of the benefit of reducing the computation period, yet keeping the annual interest unchanged, so you will negotiate an assessment period of one hour, knowing that there are **8760** hours in a year hence your deposit will grow in one year to become

$$\left(1 + \frac{1}{8760}\right)^{8760} = 2.71812$$

However, you will soon realize that there is a limit for the growth in capital deposit which is merely caused by fractioning the assessment time, and will notice that upon increased fractioning to infinitely many infinitesimal intervals, the growth of your investment (in one year) will converge to **2.71828..** times the original deposit.

This number is known as the **Euler Number "e"**

It is expressed mathematically as follows:

$$e = \lim_{n \to \infty} \left(1 + \frac{1}{n}\right)^n = 2.71828..$$

There are several mathematic and scientific applications for **e**.

It is also the base of natural logarithm. What does "*natural logarithm*" mean? It means that if **A** = e^b, then **b** is said to be the natural logarithm of **A**, or b = ln (A)

The exponential function **e**^{x} can be expressed as an infinite series:

$$e^x = 1 + \frac{x}{1!} + \frac{x^2}{2!} + \frac{x^3}{3!} + \cdots \infty,$$

which means that:

$$e = 1 + \frac{1}{1!} + \frac{1}{2!} + \frac{1}{3!} + \cdots \infty$$

e^x is the only function which equals its differentiation, that is

$$\frac{d}{dx} e^x = e^x$$

A short note: Before going back to Euler Identity I should introduce the notion **i**. "**i**" stands for "**imaginary**" and it is the square root of **-1** ($i = \sqrt{-1}$). Imaginary and complex numbers will be the subject of chapter 7 .

To prove Euler Identity we shall start with the exponential function:

$$e^x = 1 + \frac{x}{1!} + \frac{x^2}{2!} + \frac{x^3}{3!} + \cdots \infty$$

Above identity can be reached at using Taylor's expansion which was discussed earlier.

We would then substitute **x** by **ix** in the above equation, where   :
$i = \sqrt{-1}$

$$e^{ix} = 1 + \frac{ix}{1!} - \frac{x^2}{2!} - \frac{ix^3}{3!} + \frac{x^4}{4!} + \frac{ix^5}{5!} - \frac{x^6}{6!} - \frac{ix^7}{7!} + \frac{x^8}{8!} + \frac{ix^9}{9!} - \frac{x^{10}}{10!} - \frac{ix^{11}}{11!} + \frac{x^{12}}{12!} \cdots \cdots$$

$$e^{ix} = \left(1 - \frac{x^2}{2!} + \frac{x^4}{4!} - \frac{x^6}{6!} + \frac{x^8}{8!} - \frac{x^{10}}{10!} + \frac{x^{12}}{12!} \cdots\right) + i\left(x - \frac{x^3}{3!} + \frac{x^5}{5!} - \frac{x^7}{7!} + \frac{x^9}{9!} - \frac{x^{11}}{11!} \cdots\right)$$

The right hand side part of this equation is composed of two distinct bracketed expressions:

$\left(1 - \frac{x^2}{2!} + \frac{x^4}{4!} - \frac{x^6}{6!} \cdots\right)$ which is the Taylor's expansion for **cosine(x)**

as depicted in equation **III** above, and:

$$x - \frac{x^3}{3!} + \frac{x^5}{5!} - \frac{x^7}{7!} + \frac{x^9}{9!} \ ..)$$ (which is the Taylor's expansion for

sine(x) as depicted in equation **II**

Accordingly, above equation can be reduced to:

$$e^{ix} = \cos(x) + i \sin(x).$$

Substituting **x = π**, we get:

$$e^{i\pi} + 1 = 0$$ which is **Euler's Identity**

The importance of Euler's Identity is for its being the only equation that defines the relationship between the two prominent rational constants **e** and π.

In this regard; you may recall that we established – in chapter three-the relationship between π and the Golden Ratio φ (which is another prominent irrational constant), as follows

$$\pi = 10 \ \text{arcsine}(\frac{\varphi - 1}{2})$$

Remember that **arcsine(x)** - also written **tan⁻¹(x)** – means the angle whose **sine** (the **opposite / hypotenuse**) equals "**x**"

There is a couple of situations in which π appears unexpectedly in probability analysis.

First one is about picking numbers at random. It says:

If two large enough positive whole numbers are picked at random; the chance of their being relatively prime (or co-prime)

is $\dfrac{6}{\pi^2}$

For two numbers to be relatively prime, they should not divide the same prime number. If one of them is even (divisible by 2) the second should be odd and if one is divisible by 7 for example the other one should not.

Let us elaborate a little. The probability for a number picked randomly to be even is **1/2** and the probability for it to be divisible by 7 is **1/7**

In general, the probability of a number picked randomly to be divisible (without remainder) by a prime number **p**; is **1/p** and the probability for two numbers - picked randomly - to be divisible by a prime number **p** is **1 /p^2**

The probability for these two numbers not to be divisible by **p** is **(1 – 1/p^2)** which is complementary to **1/p^2** (i.e. they sum up to 1.0)

The probability for two numbers randomly selected to be co-prime is therefore:

$$P_{2ncp}= \left(1-\frac{1}{P1^2}\right)\cdot\left(1-\frac{1}{P2^2}\right)\cdot\left(1-\frac{1}{P3^2}\right)\cdot\left(1-\frac{1}{P4^2}\right)\cdot(1-\frac{1}{P5^2})\ ....$$

$$= 1/\left[\left(1-\frac{1}{P1^2}\right)\cdot\left(1-\frac{1}{P2^2}\right)\cdot\left(1-\frac{1}{P3^2}\right)\cdot\left(1-\frac{1}{P4^2}\right)\cdot(1-\frac{1}{P5^2})\ ...\right]^{-1}$$

Substitute $x = \frac{1}{p^2}$ in the identity:

$$\frac{1}{1-x} = (1-x)^{-1} = 1+x+x^2+x^3+x^5.... \infty\ -$$

which is the polynomial expansion of $(1-x)^{-1}$:

$$P_{2ncp} = 1/\left[\left(1+\frac{1}{2^2}+\frac{1}{2^4}+\frac{1}{2^6}+\frac{1}{2^8}+\cdots\right)\left(1+\frac{1}{3^2}+\frac{1}{3^4}+\frac{1}{3^6}+\frac{1}{4^8}+..\right)\right.$$

$$\left.\left(1+\frac{1}{5^2}+\frac{1}{5^4}+\frac{1}{5^6}+\frac{1}{5^8}+..\right)..\right]$$

Notice that the terms inside each bracket are the reciprocals of prime numbers raised to even powers in the form: $\dfrac{1}{(P^n)^2}$ where **p** is a prime number and **n** is any natural number.

As these terms are expanded we will get - at the denominator - all possible product combinations of the prime numbers raised to power 2.

This is nothing but the squares of all natural numbers, hence:

$$P_{2ncp} = \left(1 + \frac{1}{2^2} + \frac{1}{3^2} + \frac{1}{4^2} + \frac{1}{5^2} + \frac{1}{6^2} + .. + \infty\right)^{-1}$$

$$= \left(\frac{\pi^2}{6}\right)^{-1} = \frac{6}{\pi^2}$$

You recall Leonhard Euler's infinite series discussed earlier:

$$\frac{\pi^2}{6} = \frac{1}{1^2} + \frac{1}{2^2} + \frac{1}{3^2} + \frac{1}{4^2} \ldots \infty$$

So, the probability for two numbers picked at random to be co-prime is $\quad \dfrac{6}{\pi^2}$

You can examine this formula by selecting two numbers randomly – you will probably need the assistance of a friend of yours to ensure randomness of selection – and checking whether they are co-prime. But as the case in all probability issues, the events whose probability is examined should be numerous. A few events of number selection will have no probability substantiation value. You have to make numerous events of selecting two numbers randomly in order for your testing the formula to be viable.

The other unexpected appearance of π in a probability analysis is in the famous problem of the matchsticks and ruled sheet of paper.

This is a kind of a "***Thought Experiment***" like those discussed in chapter 2 to identify the shortest path and like others that will be discussed in chapter 10.

You will need a number of matchsticks and a sheet of paper with parallel lines spaced at the same length of a matchstick drawn on it (since it is a thought experiment, you only need to imagine having them).
You toss the matchsticks on the sheet of paper as many times as you can and take record of the number of cases in which the matchstick intersects any of the parallel lines as opposed to the total number of tossing matchsticks (see figure 6-14).

Figure 6-15 is an enlarged detail of matchstick incidence showing the two possible cases of intersection/ non-intersection with a line.

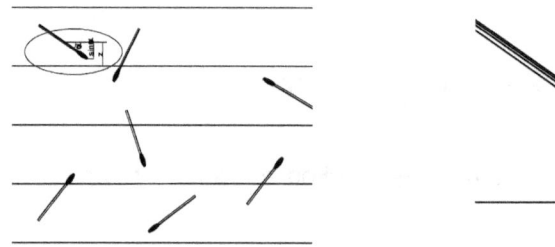

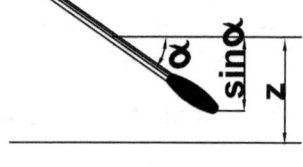

Figure 6-14

Figure 6-15

Assume that the length of a matchstick (which equals the spacing between the parallel lines) is 2 inches. Whether the matchstick crosses a line or not depends on two factors:

- The angle α which the matchstick makes with the ruled lines. This angle ranges between **o** and **π/2** (in radians)

- The distance z from the matchstick midpoint and the nearest line. This distance ranges between **0** and **1.0**

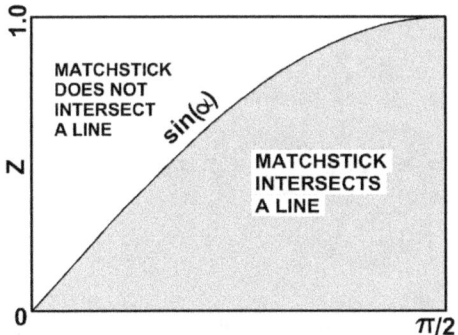

Figure 6-16

Each point in the rectangle drawn in figure 6-16 represents a unique position of the matchstick, whereby the vertical distance to the α-axis defines the distance **z** (ranging between **0** and **1**) and the horizontal distance of the point to the z-axis defines the angle **α** (in a range between **0** and **π/2**).

If the matchstick falls such that the distance **z** between its midpoint and the nearest line is smaller than sine the angle **α**, the position reference (the point in the graph representing the position) will be in the shaded area underneath the sine graph and the matchstick will intersect a line.

If otherwise the matchstick falls such that the distance **z** is greater than sine the angle **α** (the angle the matchstick makes with the lines), its position reference will be above the sine graph outside shaded area and the matchstick will not intersect any line.   The probability for the matchstick to intersect a line is the number of dots in the shaded area underneath the sine graph divided by the number of dots in the rectangle whose breadth = **π** /2 and height = **1.0**

We are not going to count the dots anyway, but should do the simpler calculation of areas.

Shaded area underneath the sine graph (representing intersections)=

$$\int_0^{\pi/2} \sin(\alpha) \, dx = -\left[\cos\left(\frac{\pi}{2}\right) - \cos(0)\right]$$

$$= 1.0$$

Area of the whole rectangle containing all intersection possibilities (both intersecting and non-intersecting)

$$= \left(\frac{\pi}{2}\right) * 1.0 = \frac{\pi}{2}$$

The probability for matchstick to cross a line = (shaded area) / the whole area of the rectangle =

$$\frac{1}{\left(\frac{\pi}{2}\right)} = \frac{2}{\pi}$$

The Chapter Quiz

π (the circle's circumference to diameter ratio), φ (the Golden Ratio) and **e** (Euler's constant) are the most important mathematical constants. The three of them are irrational.

In chapter three, a relationship between π, the circle's circumference to diameter ratio and the Golden ratio φ was established as follows

$$\pi = 10 \arcsin\left(\frac{\varphi - 1}{2}\right)$$

In this chapter; we have seen the amazing Euler's Identity which defined the relationship between Euler Constant "**e**" and "π" as follows:

$$e^{i\pi} + 1 = 0$$

From these two formulae; find a relationship between **e** and φ then demonstrate in a graphical presentation the relationship between each two of these three constants

Discussion and solution of the Chapter Quiz

$e^{i\pi} + 1 = 0$ (Euler's identity), hence $e^{i\pi} = -1$

Raise both sides to a power of **- i**

$$e^{\pi} = -1^{-i}$$

In the above equation, substitute

$$\pi = 10\ arcsine((\varphi/2 - 1)/2)$$

$$e^{10\ \text{arcsine}((f-1)/2)} = (-1)^{-i}$$

A graphical presentation the relationship between each two of these three constants is shown in figure 6-17

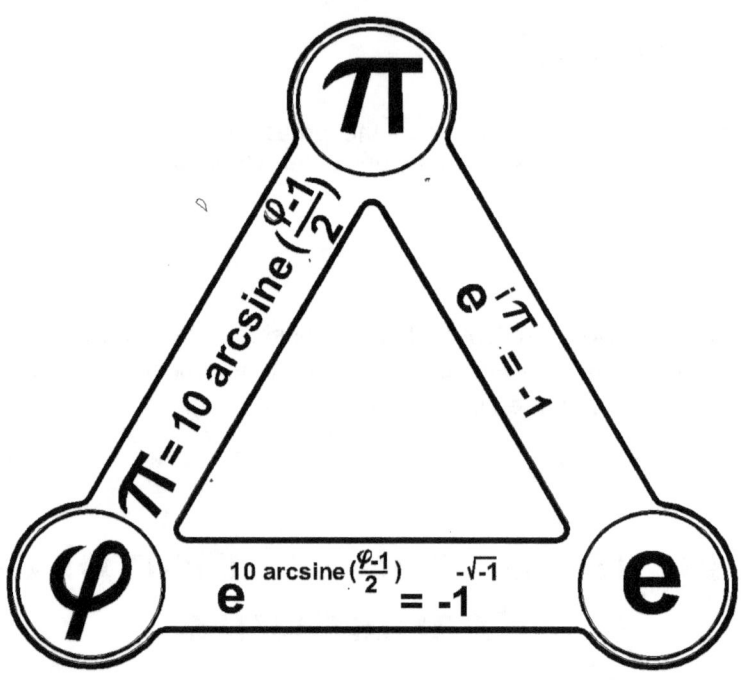

Figure 6-17

Chapter 7:   Prime Numbers

A prime number is a whole number greater than 1 which is only divisible by 1 and by itself such as **2, 3, 5, 7, 11, 13, 17, 19, 23, 29** .. Etc.

Other numbers are called composite because they can be factored into prime numbers.

That is why prime numbers have always been considered as the building blocks of the entire set of whole numbers and the most basic of it.

Prime numbers have occupied a prime position in the study of number theory. Many questions related to prime numbers have been intriguing mathematicians for centuries, and some of these questions are still waiting for answer.

The distribution of prime numbers within the set of whole numbers is an example.

GENERATING PRIMES

Prime numbers appear to scatter in an unpredictable manner. Exploring their distribution and searching for a mathematical pattern or a formula to generate them has long been the quest of several theory researchers.

For example; a sequence like **31, 331, 3331, 33331, 333331, 3333331, 33333331** looks promising.

Numbers in that sequence obviously pass the tests of divisibility by **2, 3, 5, 7** and **11** Divisibility by these numbers was discussed elaborately in chapter 5, so let us review divisibility status in light of the guidelines provided therein:

- **33...3331** is not divisible by **2** because the units digit is an odd number,

- **33...3331** is not divisible by **3** because the sum of its digits isn't

- **33...3331** is not divisible by **5** because the units digit is neither **0** nor **5**

- Applying the test of divisibility by **7** as discussed in chapter 5 on **33331** will reduce it to **3333 – 2*1 = 3331**. Applying the same process iteratively will reduce it further to **331**; then **31** which is not divisible by **7**.

- **33...3331** is not divisible by **11** because the difference between oddly ordered digits and evenly ordered digits S_{ODD} - S_{EVEN} = **1** or **2**; none of which is divisible by 11

This was encouraging and raised the aspirations that such a pattern can generate prime numbers. Yet, the pattern fails the test at next term since **333333331 = 17*19607843**

Some formulas to generate prime numbers were initially thought to be working, yet they eventually failed the test at a certain point.

One famous formula is $X^2 + X + 41$.

Substituting X by any number between **1** and **39** would yield a prime number as shown in the tabulation in figure 7-1

However, for **x** = 40 and 41, $X^2 + X + 41$ = 1681 and **1763**, both of which are composite numbers that have **41** as a factor.

This is expected any way because for **X = 40**; $X^2 + X + 41$ = **40*40 + 40 + 41**

= **40(40+1) + 41 = 41²**. And for **X = 41**; $X^2 + X + 41$ = **41(41 + 1 + 1) = 41*43**

EXAMINING WHETHER THE FORMULA							$X^2 + X + 17$						
GENERATES PRIME NUMBERS (FOR A RANGE X= 1 TO 42)													
X	1	2	3	4	5	6	7	8	9	10	11	12	13	14
X² + X + 17	19	23	29	37	47	59	73	89	107	127	149	173	199	227
PRIME?	Y	Y	Y	Y	Y	Y	Y	Y	Y	Y	Y	Y	Y	Y
X	15	16	17	18	19	20	21	22	23	24	25	26	27	28
X2 + X + 17	257	289	323	359	397	437	479	523	569	617	667	719	773	829
PRIME?	Y	N	N	Y	N	N	Y	Y	Y	Y	N	Y	Y	Y
X	29	30	31	32	33	34	35	36	37	38	39	40	41	42
X2 + X + 17	887	947	1009	1073	1139	1207	1277	1349	1423	1499	1577	1657	1739	1823
PRIME?	Y	Y	Y	N	N	N	Y	N	Y	Y	N	Y	N	Y

Figure 7-1

Another formula that was discovered by the Swiss mathematician Leonhard Euler in the eighteenth century is: $X^2 + X + 17$ which generates prime numbers for discrete ranges of X= 1 to 15, but fails repeatedly at X > 15 as shown in the tabulation at figure 7-2.

EXAMINING WHETHER THE FORMULA $X^2 + X + 41$														
GENERATES PRIME NUMBERS (FOR A RANGE X= 1 TO 42)														
X	1	2	3	4	5	6	7	8	9	10	11	12	13	14
$X^2 + X + 41$	43	47	53	61	71	83	97	113	131	151	173	197	223	251
PRIME?	Y	Y	Y	Y	Y	Y	Y	Y	Y	Y	Y	Y	Y	Y
X	15	16	17	18	19	20	21	22	23	24	25	26	27	28
X2 + X + 41	281	313	347	383	421	461	503	547	593	641	691	743	797	853
PRIME?	Y	Y	Y	Y	Y	Y	Y	Y	Y	Y	Y	Y	Y	Y
X	29	30	31	32	33	34	35	36	37	38	39	40	41	42
X2 + X + 41	911	971	1033	1097	1163	1231	1301	1373	1447	1523	1601	1681	1763	1847
PRIME?	Y	Y	Y	Y	Y	Y	Y	Y	Y	Y	Y	N	N	Y

Figure 7-2

In the seventeenth century; the French philosopher and mathematician Marin Mersenne (1588 – 1648) studied large prime numbers.

He used the following formula - which was later named after him - to generate primes:

$M_P = 2^P - 1$

where **P** is a smaller prime number which is processed by the formula to generate the larger prime M_P

For example; for **P = 5**, $M_p = 2^5 - 1$ = 31 and

For **P = 7**, $M_p = 2^7 - 1 = 128 - 1$ = 127

However; the formula doesn't hold long. It fails at **P = 11:**

$M_{11} = 2^{11} - 1 = 2048 - 1 = 2047$ which is not prime (it is the product of 23 & 89)

Similarly, M_{23} and M_{29} failed the test. They aren't primes.

It was eventually perceived that none of the expected patterns or devised formulas could generate prime numbers.

The only reliable method to separate prime numbers in a given range of natural numbers is that known as **Eratosthenes Sieve** so named after its developer the Greek mathematician **Eratosthenes of Cyrene** (276 – 194 BC).

In Eratosthenes Sieve, natural numbers in a certain range starting from 1 – say **1** to **N** - are listed then starting with **2** as the first prime, all its multiples (even numbers) are stricken off. Multiples of **3** are then stricken off, followed by multiples of **5** then multiples of **7** and so on until the same is done with the multiples of the prime **Pₑ** where

$$P_e > \sqrt{N}$$

All other numbers in the list which would be left un-struck are the primes falling within the range **1** to **N**

Eratosthenes sieve for the first 200 natural numbers is shown in figure 7-3

1	2	3	4	5	6	7	8	9	10	11	12
13	14	15	16	17	18	19	20	21	22	23	24
25	26	27	28	29	30	31	32	33	34	35	36
37	38	39	40	41	42	43	44	45	46	47	48
49	50	51	52	53	54	55	56	57	58	59	60
61	62	63	64	65	66	67	68	69	70	71	72
73	74	75	76	77	78	79	80	81	82	83	84
85	86	87	88	89	90	91	92	93	94	95	96
97	98	99	100	101	102	103	104	105	106	107	108
109	110	111	112	113	114	115	116	117	118	119	120
121	122	123	124	125	126	127	128	129	130	131	132
133	134	135	136	137	138	139	140	141	142	143	144
145	146	147	148	149	150	151	152	153	154	155	156
157	158	159	160	161	162	163	164	165	166	167	168
169	170	171	172	173	174	175	176	177	178	179	180
181	182	183	184	185	186	187	188	189	190	191	192
193	194	195	196	197	198	199	200	201	202		

Figure 7-3:   Eratosthenes sieve

The sieve lists the following first **46** primes:

2	3	5	7	11	13	17	19	23	29	31	37	41
43	47	53	59	61	67	71	73	79	83	89	97	
101	103	107	109	113	127	131	137	139	149	151	157	
163	167	173	179	181	191	193	197	199				

THE DISTRIBUTION OF PRIMES

The distribution of primes within a set of natural numbers has long attracted the attention of mathematicians.

It has been noticed that the intensity of primes within a range 1 to N of natural numbers is reduced as N increases.

As shown in figure 7-3; there are **25** primes in the **first hundred** natural number and **21** in the **second hundred**.

In a larger scale, while there are **95** primes in the first **500** natural number, there are only **73** primes in the second **500** natural number.

The trend is obvious, the intensity of prime numbers is reduced as the lower and upper bounds of the range are increased.

Such continual decrease in intensity made some of ancient mathematicians think that the intensity will keep decreasing as the range goes up such that beyond a certain point; primes will cease to exist.

Euclid dismissed this concept and gave an elegant proof – which will be discussed shortly - that primes would continue to emerge infinitely

A theorem assessing the average distribution of primes among other numbers was developed by the end of 19th century by the French mathematician Hadamard and Belgian mathematician Poussin conjures that :

$$\frac{\text{Number of primes within a range 1 to N}}{N} \sim \frac{1}{\ln(N)}$$

(The sign $\sim$ means approximately equals)

Let us test the theorem for **N = 500** and **1000** and let **Np** be number of primes within a range of natural numbers from **1** to **N**

N	Np	Np/N	1 /ln(N)
500	95	0.19	0.161
1000	168	0.168	0.145

The equation **Np/ N** ~ **1/ ln(N)** does not yield an impressive approximation for **N** is as low as **1000**.

The intensity of primes within a range of 1 to N fades out as N gets lager. This triggered the question: *Do we have a limiting number beyond which they cease to exist?*

Euclid made a simple and elegant proof by contradiction on the infinitude of prime numbers.

The proof initially assumes that there is a finite number of primes and that P_n is the largest and the last prime beyond which no primes exist. Euclid then considered the expression **N**:

N = (2 * 3 * 5 * 7 * 11 * 13 * 17 * 19 * 23 * 29 * .... * P_n) + 1

The amount between brackets is obviously a composite number which is the product of all prime numbers that exist; hence it is divisible by all these prime numbers.

N is not therefore divisible by any prime number and there will always be a remainder of 1 in any such division.

Accordingly, **N** is only divisible by itself and by 1 hence it is a prime number and it is obviously greater than P_n.

This result contradicts the initial assumption that P_n is the largest and the last prime beyond which no primes exist, and proves the infinitude of primes.

The Chapter Quiz

Fermat's Little Theorem can help in testing whether a given number is prime.

The theorem was given the name of its discoverer Pierre de Fermat (1601 – 1665), whose name was also given to Fermat Last Theorem which was briefly discussed in Chapter One.

FERMAT'S LITTLE THEOREM

The Theorem

Let "p" be prime and "a" is any integer which is co-prime with "p"

$(a^{P-1} - 1)$ is divisible by **p**

In modular arithmetic notation (you may refer to chapter 5 for elaboration):

$$a^{P-1} \equiv 1 \bmod p$$

And $(a^P - a)$ is also divisible by **p**

In modular arithmetic notation:

$$a^P \equiv a \bmod p$$

Elaborate on and prove Fermat's Little Theorem

Discussion and solution of the Chapter Quiz

Fermat's Little Theorem states that:

If "**p**" is a prime number and a is a positive integer, $(a^P - a)$ is divisible by **p**

"**p**" should be prime for Fermat's Little Theorem to remain valid.

Let us check – using a few examples – what will happen if we deviate from this condition.

Let p = 5 (*prime*) and a = 6, $(a^P - a)$ = 7776 – 6 = 7770

(divisible by p = 5)

Let p = 7 (*prime*) and a = 4, $(a^P - a)$ = 16384 – 4 = 16380

(divisible by p = 7)

Let p = 3 (*prime*) and a = 6, $(a^P - a)$ = 216 – 6 = 210

(divisible by p = 3)

Let p = 2 (*prime*) and a = 5, $(a^P - a)$ = 25 – 5 = 20

(divisible by p = 2)

Now to verify by contradiction, let us see what will happen if **p** is a **nonprime** number:

Let p = 9 (nonprime) and a = 2, $(a^P - a)$ = 512 – 2 = 510

not divisible by p = 9

Let p = 8 (nonprime) and a = 3, $(a^P - a)$ = 6561 – 3 = 6558

(not divisible by p = 8)

Let p = 6 (nonprime) and a = 5, $(a^P - a)$ = 15625 – 5 = 15620

(not divisible by p = 6)

Let p = 6 (nonprime) and a = 8, $(a^P - a)$ = 262144 – 8 = 262136

(not divisible by p = 6)

THE PROOF

1. Applying the Modular Arithmetic principles

Each of the numbers between **1** and **p-1** will have *different mod p*, meaning that you will not find among these **p-1** numbers two modularly congruent numbers (the remainder of the division by **p** is different in each number).

Let us take the numerical example of **p** = 7.

The numbers **1, 2, 3, 4, 5 & 6** will have *different mod 7* .

The remainder of the division **1/7** is **1**, the remainder of the division **2/7** is **2** and the remainder of the division **3/7** is **3** and so on.

If each of these **6** numbers is multiplied by another number which is **co-prime with 7** – say **10** – we will get the following six numbers **10, 20, 30, 40, 50 & 60** which also have *different mod 7.*

Furthermore, the product of these numbers will not be divisible by **7**.

Consider the **p -1** numbers **a, 2a, 3a, 4a, ... (p - 2)a, (p - 1)a**.

These numbers are *different mod p*. The product of these **(p - 1)** numbers is:

(p – 1)! a^{p-1}

where **(p – 1)!** means **factorial (p – 1)** = **(p - 1)(p - 2)(p - 3) ... 3*2*1**

Since **a** is co-prime to **p**, the product of:
$[a, 2a, 3a, 4a .... (p-1)a]$ must be congruent modulo *p* to the product of $[1, 2, 3, 4 .... (p-1)]$

Why is that?

Because - if p and a are relatively co-prime - the remainders of the divisions

$$\frac{a}{p}, \frac{2a}{p}, \frac{3a}{p}, \frac{4a}{p}, \dots \frac{(p-1)a}{p} \text{ are } 1, 2, 3, 4 \dots (p\text{-}1)$$

but not necessarily in the same order, so these remainders will simply need a positional rearrangement - if sorting them to an ascending order matters.

However, since we are only concerned with the product of multiplication, we shouldn't bother about the order of arranging the remainders.

An explanation is provided in the examples of figure 7- 4

Hence: **$(p - 1)! \cdot a^{p-1} \equiv (p - 1)! \bmod p$**

Dividing the two sides of the above equation by **$(p - 1)!$** we get:

$a^{p-1} \equiv 1 \bmod p$, hence

$(a^{p-1} - 1)$ is divisible by **p**

Multiplying this expression by **a** we get:

$(a+^p - p)$ is also divisible by **p**

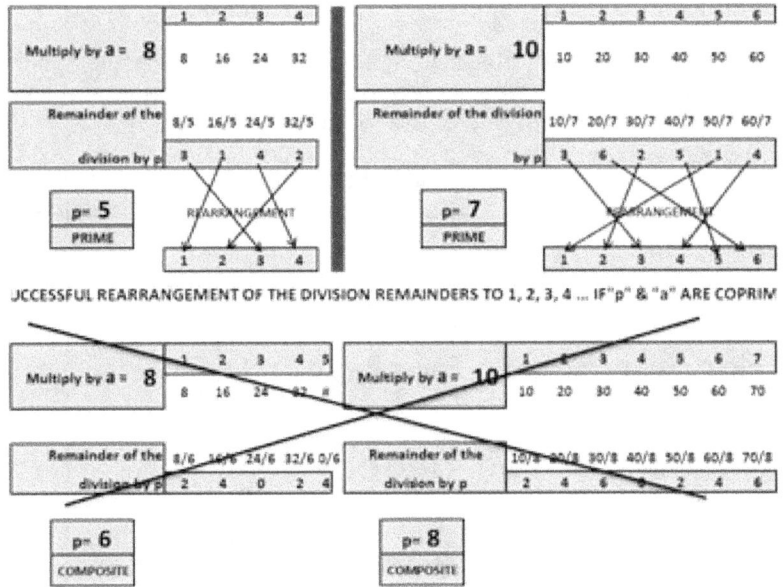

Figure 7-4

Theorem:

If p and a are relatively co-prime, the remainders of the divisions

$$\frac{a}{p}, \frac{2a}{p}, \frac{3a}{p}, \frac{4a}{p}, \ldots \frac{(p-1)a}{p} \quad \text{are } 1, 2, 3, 4 \ldots (p-1)$$

(with some positional rearrangement)

2.   Combinational proof

The technique used in this proof is not traditional. It is based on composing graphical combinations.

Assume you have beautiful beads of "**a**" different colors, and that you want to thread strings of these beads with "**p**" beads in each thread.

You will join the ends of these strings to form closed bracelets. Now you have a number of bracelets, each one holds "**p**" beads and each one of the beads is given one of the "**a**" colors used.

You will have all possible combinations of arranging beads with different colors in threads and after that you will group bracelets of similar patterns.

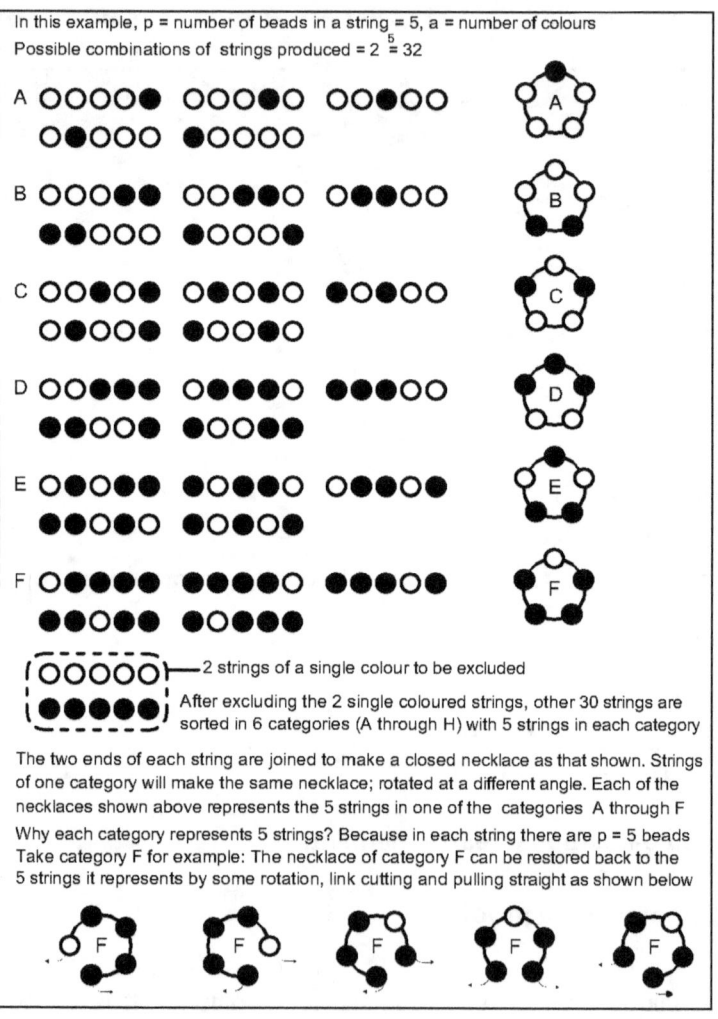

In this example, p = number of beads in a string = 5, a = number of colours
Possible combinations of strings produced = 2^5 = 32

2 strings of a single colour to be excluded

After excluding the 2 single coloured strings, other 30 strings are sorted in 6 categories (A through H) with 5 strings in each category

The two ends of each string are joined to make a closed necklace as that shown. Strings of one category will make the same necklace; rotated at a different angle. Each of the necklaces shown above represents the 5 strings in one of the categories A through F

Why each category represents 5 strings? Because in each string there are p = 5 beads
Take category F for example: The necklace of category F can be restored back to the 5 strings it represents by some rotation, link cutting and pulling straight as shown below

Figure 7-5

While grouping; you will be allowed to rotate any bracelet for its pattern to be matching that of another one.

Well, let us see such amusing bracelet making in the example of figure 7-5

In the example of figure 7-5, number of beads in a string p = **5**, and number of colors available **a** = **2**.
To calculate all possible design combinations of strings consider the possibilities in each bead position. There are two possibilities for the first bead. It will have to be either one of the two colors available (number of colors **a** = **2**).

The same applies to second, third, fourth and fifth beads, so the total number of combination = **2 x 2 x 2 x 2 x 2 = 2^5 = 32** possible combinations which is the number of strings in the example of figure 7-5.

Some of these strings contain **5** beads of the same color.

Number of these mono-colored strings is the number of colors: **a** = **2**.

Strings of one category will form bracelets of the same design after connecting the ends of each string and rotating it to match other bracelets in the category.

Matching process is explained in the example of figure 7-5.

By excluding the "**a**" strings (those who have a single color), remaining **30** strings can be sorted under **6** categories (**A** through **F**), with each category including **5** strings (**p**)

The mathematics underlying these combinations is:

- Number of strings; each formed of p = **5** beads with each bead having one of **a** = **2** colors is: 2^5 = **32** strings

- If the **a** = **2** strings having a single color are excluded, balance strings can be sorted under **6** categories with each category representing p = **5** +strings that make the same design of a bracelet (after some rotation)

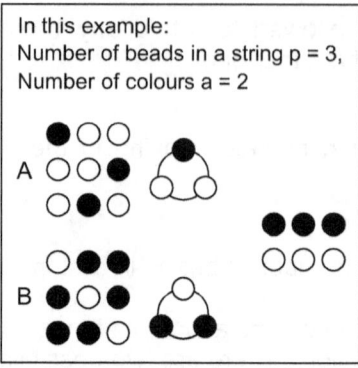

In this example:
Number of beads in a string p = 3,
Number of colours a = 2

Figure 7-6

Bottom line is that $2^5 - 2 = 6 * 5$ or $(a^p - p)$ = multiple of p. This proves that $(a^p - p)$ is divisible by p

Another simpler example is presented in figure 7-6 in which number of colors "**a = 2**" (also) and number of beads in a string "**p = 3**"

Number of possible combinations (number of strings) = $a^p = 2^3 = 8$, **two** of which **(a)** are excluded for being of a single color

The other **6** strings are sorted under **2** categories each one represents **p** = 3 strings. Each of these 3 strings makes a bracelet of the same design (after some rotations)

Bottom line is that $2^3 - 2 = 2 * 3$ or $(a^p - p)$ = multiple of p

This further confirms that $(a^p - p)$ is divisible by **p**

We still have to show that for this modularity relationship to remain valid; **a** and **p** should be relatively co-prime.

What will happen if the number of beads in a string **p** and the number of colors **a** are not co-prime to each other. For example if **p** is taken **4** and **a** taken **2** ?

A string formed of **4** beads like that one ●○●○ is actually composed of two identical shorter substrings having **2** beads each

like that one ●○

We remember that when the number of beads in a string "**p**" was prime such splitting of the string to shorter substrings was not possible, and that the number of strings under the same bracelet category was equal = **p (4 in this case)**.

We shou+ld not expect that the above 4-bead string will behave in the same way. Combinations in that kind of strings (formed of identical substrings) will correspond to the combinations of the smaller substring which contains two beads only.

This means that the $(a^p - p)$ combinations (strings) will not fit in bracelet categories of equal number of strings, because a category representing that kind of strings will contain a smaller number of strings than **p**.

An example for a combinatory case with "**p**" being non-prime is shown in figure 7-7

We remember that when the number of beads in a string "**p**" was prime such splitting of the string to shorter substrings was not possible, and that the number of strings under the same bracelet category was equal = **p (4 in this case)**.

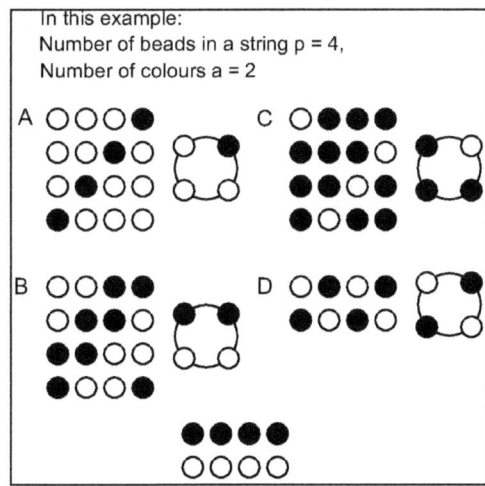

Figure 7-7

We should not expect that the above 4-bead string will behave in the same way. Combinations in that kind of strings (formed of identical substrings) will correspond to the combinations of the smaller substring which contains two beads only.

This means that the $(a^p - p)$ combinations (strings) will not fit in bracelet categories of equal number of strings, because a category representing that kind of strings will contain a smaller number of ++strings than **p**.

An example for a combinatory case with "**p**" being non-prime is shown in figure 7-7

Chapter 8: Heron's Formula

Heron of Alexandria (10 – 70 AD) is a Greek engineer and mathematician who was born and lived in Alexandria, Egypt.

Heron's formula calculates the area of a triangle given the length of its three sides.

This is a much more practical area calculation method as opposed to the common method requiring the measure of an altitude.

Unlike quadrilaterals, pentagons and other geometric entities having more than three sides, a triangle is uniquely identified if the lengths of its three sides are given.

That is why triangulation (a surveying method based on dividing a piece of land into triangles) is a very prominent land surveying tool and that is why Heron's formula is quite important in supporting triangulation.

The formula constitutes a great tool for engineers and surveyors in computing territorial areas and in setting out plot plans and maps.

However, this was not the only reason for me to present Heron's formula here! Another important reason is that the derivation of the formula and the proofing strategy constitute an impressive model of mathematical reasoning characterized by deep insight, smart algorithm and unexpected but well thought of moves.

Dunham (1990) simulates such moves with actions in Agatha Christie's thriller novels in which you would be a few pages away from the end yet you still have no clue on where would the events be driven to

Given are the measure of sides **a**, **b** and **c** of a triangle as that shown in figure 8-1

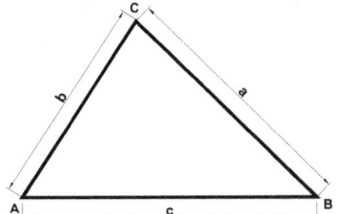

Figure 8-1

THE FORMULA

The formula derived by Heron is:

Area of triangle:

$$A_\triangle = \sqrt{s(s-a)(s-b)(s-c)}$$

Where **s is the semi-perimeter = triangle's perimeter / 2** and **a, b** & **c** are the side lengths of the triangle

First step in the proof trip programmed by Heron is to draw the circle inscribed in triangle ABC. The center "O" of the circle is called the **INCENTER** of triangle ABC.

This **INCENTER** is the meeting point of the bisectors of the interior angles A, B and C of the triangle ABC.

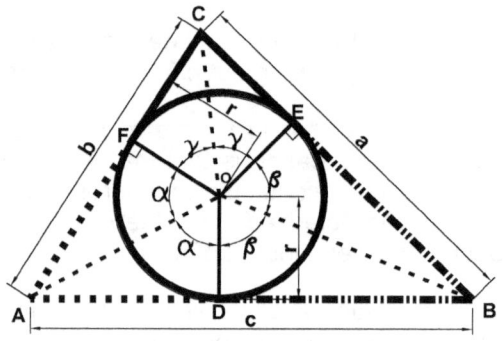

Figure 8-2

Triangle ABC with the inscribed circle and the INCENTER "O" are shown in figure 8-2

Notice that line segments of equal length in figure 8-2 are given the same line style being *continuous*, *dashed* or *dash-dots*. Notice also that lines extended from A, B and C to the INCENTER "O" have split the triangle **ABC** into three triangles:

- **BOC, the area of which = a . r /2**
- **COA, the area of which = b . r /2**
- **AOB, the area of which = c . r /2**

where **"r"** is the radius of inscribed circle

The area of triangle **ABC** (which is required by this analysis) equals the sum of areas of above 3 triangles, hence:

Area of △ ABC = (a + b + c). r/2 = (the triangle's perimeter) * **r /2.** In other words:

    Area of △ ABC = s* r   .... **(1)**

Where **"s"** is the triangle's semi-perimeter and **"r"** is the radius of inscribed circle

Take also another notice that those lines extended from **A**, **B** and **C** to the INCENTER "**O**" have created three pairs of congruent triangles:

- **A pair of congruent triangles AFO & ADO having dotted outer edges and interior angles α , 90, and (90 – α)**

- **A pair of congruent triangles DBO & EBO having dash-dot outer edges and interior angles β , 90, and (90 – β)**

- **A pair of congruent triangles CFO & CEO having continuous outer edge lines and interior angles γ , 90, and (90 – γ)**

Now watch the process of Heron's genius proof. You will only know the motivations of these following steps at the end of the trip. However, one strategic goal can be seen throughout. Apparently, Heron wanted to gather all key elements in one line to gain better control over the proof process.

This line is the base of the triangle **AB** and the extension that will be added to it. Heron decided to extend **BA** to **G** such that **AG** will be equal to **FC**

Furthermore, Heron decided to construct line **OH** perpendicular to **OB** and line **AH** perpendicular to **AB** (see figure 8-3).

Lines **OH** and **AH** will obviously meet at **H**.

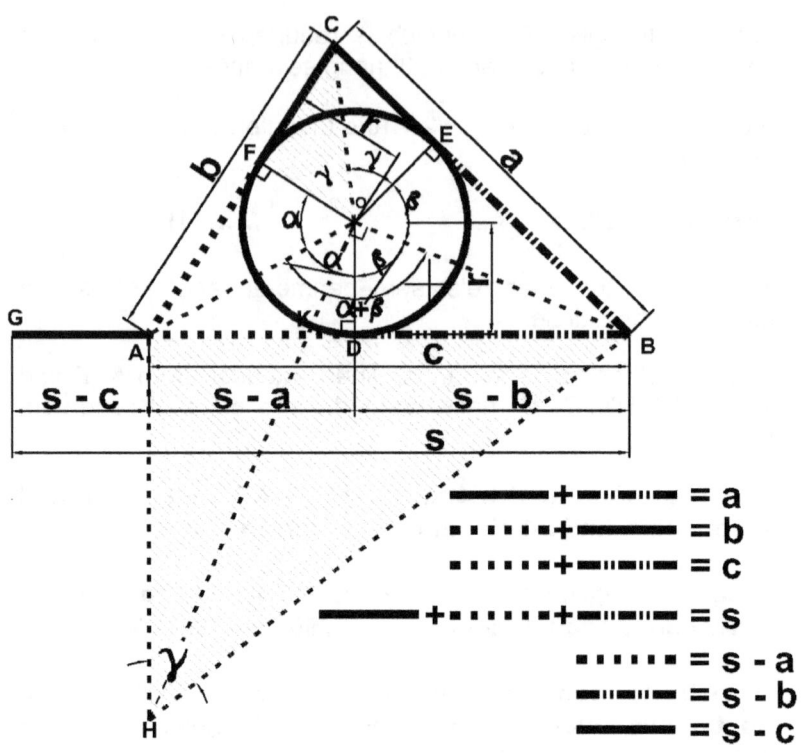

Figure 8-3: A key figure in Heron's proof

You will notice – in figure 8-3 – that Heron smartly assigned line segment to represent the key elements in his formula: **s**, **(s-a)**, **(s-b)** and **(s-c)**; along the triangle base line **GB:**

s = GB , (s – a) = AD , (s – b) = BD , and   (s – c) = AG   ... (2)

Next step in Heron's marathon towards proof is to establish similarity of the triangles AHB and FOC (both shaded in figure 8-3).

To achieve that, he had to prove the congruence of angles AHB and FOC. Quadrilateral **OAHB** (taken from +figure 8-3) is being presented separately as figure 8-4

You may recall that in the discussion of Chapter One quiz; it was established that a triangle drawn in a semicircle with its base coinciding with the diameter; would be a right angled triangle.

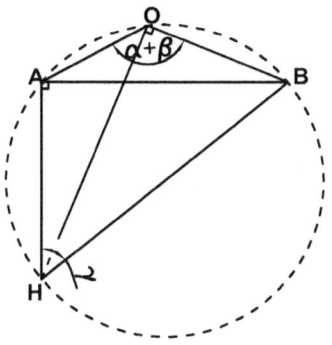

Figure 8-4

The reverse of this postulate is valid. A right angled triangle is uniquely circumscribed by a circle in which the hypotenuse will be a diameter. This means that triangles HAB and HOB – in figure 8-4 - can be uniquely circumscribed by circle NAOB in which the diameter HB is the common hypotenuse of the two triangles. It will naturally follow that AHBO is a cyclic quadrilateral.

You may further recall that in the discussion of Chapter Two quiz; it was established that: "*Each two opposite interior angles in the quadrilateral sum up to 180° *". This means that angle **AHB** is supplementary to angle **BOA** (figure 8-4). Angles meeting at point "O" of figure 8-3 and summing to **360°** are **2** α , **2** β and **2** γ, hence $\alpha + \beta + \gamma$ **= 180°**

In figure 8-4, ∠**AOB** + ∠**AHB** = 180o as established, hence

∠**AOB** + ∠**AHB** = angles $\alpha + \beta + \gamma$, but ∠**AOB** = ∠α + ∠β as clearly shown in figures 8-3 and 8-4, hence angle **AHB** = angle γ

The aim of this preceding argument is to establish the similarity of triangles **AHB** and **FOC** in figure 8-3. The two triangles are also drawn separately in the right-hand side of figure 8-5 below.

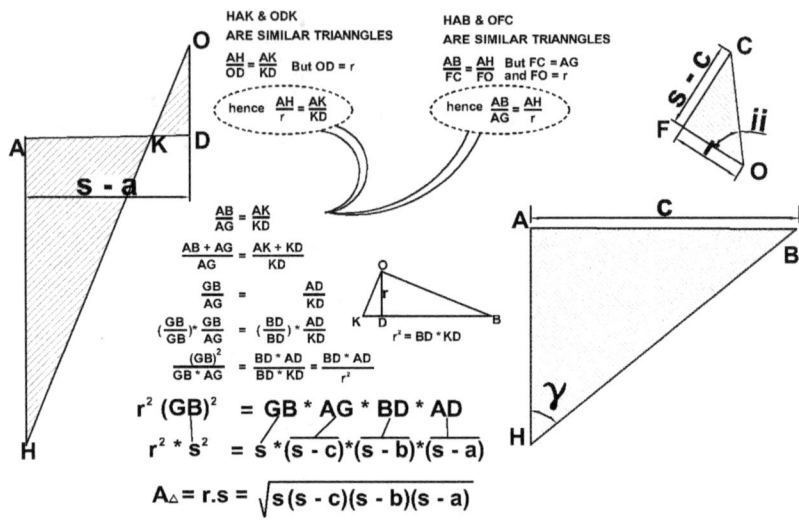

Figure 8-5

Why triangles **"AHB"** and **"FOC"** (right-hand side of figure 8-5) are similar?

Angle **AHB** in triangle **AHB** = angle **FOC** in triangle **FOC** = γ (just proven) and

Angle **HAB** in triangle **AHB** = angle **CFO** in triangle **FOC** = **90°** (given)

RESULT OF SIMILARITY

$$\frac{AB}{CF} = \frac{AH}{FO}$$ but **CF = AG** and **FO = r** (see figure 8-3),

hence

$$\frac{AB}{AG} = \frac{AH}{r} \qquad\qquad \text{...... (3)}$$

Now consider the similarity of triangles KAH and KDO (left-hand side of figure 8-5)

Why triangles **"KAH"** and **"KDO"** are similar?

Angle **HKA** in triangle **KAH** = angle **OKD** in triangle **KDO** (opposite angles of two intersecting lines)

And angle **HAK** in triangle **KAH** = angle **ODK** in triangle **KDO** = **90°** (given)

RESULT OF SIMILARITY

$$\frac{AH}{OD} = \frac{AK}{KD}, \text{ but } \mathbf{OD = r} \text{ (see figure 8-3)},$$

hence $\dfrac{AH}{r} = \dfrac{AK}{KD}$ $\qquad$ ....(4)

But $\dfrac{AB}{AG} = \dfrac{AH}{r}$ - from **(3)** – hence $\dfrac{AB}{AG} = \dfrac{AK}{KD}$

Adding 1 to each side of this latter equation:

$$\frac{AB}{AG} + 1 = \frac{AK}{KD} + 1$$

Therefore $\dfrac{AB+AG}{AG} = \dfrac{AK+KD}{KD}$

hence $\dfrac{GB}{AG} = \dfrac{AD}{KD}$

(see figure 8-3).

Multiply left side by **(GB/GB)** and right side by **(BD/BD)**

$$\left(\frac{GB}{GB}\right) * \frac{GB}{AG} = \left(\frac{BD}{BD}\right) \frac{AD}{KD}$$

Hence
$$\frac{GB^2}{GB*AG} = \frac{BD*AD}{BD*KD} \qquad\qquad ....(5)$$

Triangle OKB and its altitude OD have been extracted from figure 8-3 and displayed below separately as figure 8-6 to consider a third case of triangular similarity.

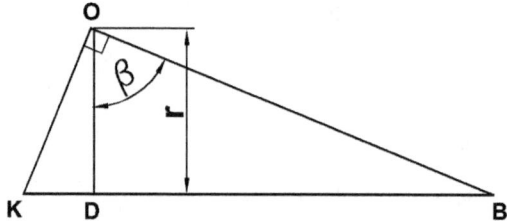

Figure 8-6

Triangles ODK and BDO are similar because ∠ DOK = ∠ DBO (both complementary to ∠ OKD) and ∠ ODK = ∠BDO = **90°** , hence:

$$\frac{KD}{OD} = \frac{OD}{BD} \quad \text{from which}$$

BD.KD = OD^2 = r^2 ....(6)

Substituting "**BD.KD**" in equation **(4)** by "**r^2** " **(equation 5)** we get:

$$\frac{GB^2}{GB * AG} = \frac{BD * AD}{r^2} \quad\quad ....(7)$$

$$r^2 . (GB)^2 = GB * AG * BD * AD \quad .... \quad (8)$$

From (1) we have:

s = GB , (s – a) = AD , (s – b) = BD and (s – c) = AG [see (2)]

From (**8**) and (**2**) we get:

$$r^2 * s^2 \quad = s * (s – a) * (s – b) * (s – c) .... \quad (9)$$

From (1), $A_\Delta = s * r$, substituting in (**9**) we get:

$(A_\Delta)^2 \quad = s * (s – a) * (s – b) * (s – c)$, hence

$$A_\Delta = \sqrt{s(s-a)(s-b)(s-c)}$$

And here we reach the end of the proof journey.

The Chapter Quiz

Prove the **Pythagorean Theorem** using **Heron's formula**

Discussion and solution of the Chapter Quiz

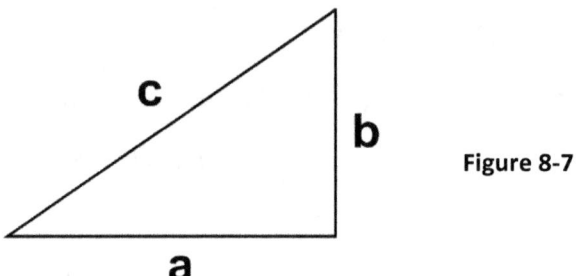

Figure 8-7

Let **A** be the area of the triangle in figure 8-7

Heron's formula states that:

$$A = \sqrt{s(s-a)(s-b)(s-c)}$$
$$A^2 = s(s-a)(s-b)(s-c)$$

The semi-perimeter $s = \dfrac{1}{2}(a+b+c)$

$$s-a = \frac{1}{2}(-a+b+c)$$
$$s-b = \frac{1}{2}(a-b+c)$$
$$s-c = \frac{1}{2}(a+b-c)$$

Substituting the above expressions of (s − a), (s − b) and (s − c) in the equation $A^2 = s(s-a)(s-b)(s-c)$ we get:

$$A^2 = \tfrac{1}{2}(a + b + c)* \tfrac{1}{2}(-a + b + c) * \tfrac{1}{2}(a - b + c) * \tfrac{1}{2}(a + b - c)$$

Multiplying the equation's two sides by **16**

$$16\,A^2 = (a + b + c)* (-a + b + c)* (a - b + c)* (a + b - c)$$

The area **A** also equals half the product of its base and its altitude (**a.b/ 2**). Substituting in above equation we get:

$$4\,a^2 b^2 = (a + b + c) * (-a + b + c) * (a - b + c) * (a + b - c)$$

$$= (2\,a^2 b^2 + 2\,b^2 c^2 + 2\,c^2 a^2 - a^4 - b^4 - c^4)$$

$$a^4 + 2a^2 b^2 + b^4 = 2(b^2 c^2 + c^2 a^2) - c^4$$

$$(a^2 + b^2)^2 = 2\,c^2(a^2 + b^2) - c^4$$

$$(a^2 + b^2)^2 - 2\,c^2(a^2 + b^2) + c^4 = 0$$

$$[(a^2 + b^2) - c^2]^2 = 0$$

$$(a^2 + b^2) - c^2 = 0, \text{ hence}$$

$$a^2 + b^2 = c^2$$

And this proves the Pythagorean Theorem

Chapter 9: Imaginary and Complex Numbers

THE SQUARE ROOT OF A NEGATIVE NUMBER

Let us start with a simple riddle:

Can you split the number 8 into two numbers the product of which is 15? You will find - in a short time - that the answer of the puzzle is 3 and 5

But what if the puzzle question was: Split the number 8 into two numbers the product of which is 150? Let us try to solve it algebraically:

Let these two numbers be "**x**" and "**8 − x**". The product of the two numbers is: **x (8 − x) = 150**, hence **x² − 8x + 150 = 0**. This is a quadratic equation in the form "**ax² + bx + c = 0**" for which the solution is:

$$x = \frac{-b \pm \sqrt{b^2 - 4ac}}{2a} = \frac{8 \pm \sqrt{8^2 - 4*1*150}}{2}$$
$$= 4 \pm \sqrt{-134}$$

so the two numbers that sum up **8** and have a product of **150** are:

$$4 + \sqrt{-134}) \text{ and } (4 - \sqrt{-134})$$

But what does the square root of a negative number like **− 134** mean? We know that the square of **5** is **25** and that the square of -5 is also 25, so the square of any real number (whether positive or negative) is always a positive number because **(-x)² = x²**

The square roots of negative numbers kept emerging in algebraic calculations as such without gaining any recognition by the mathematical community, much the same as happened with the negative numbers on their early appearance.

The words of the 12th century Indian mathematician **Bhaskara** (1114 - 1185) on that matter are being quoted here: "***The square of any*** *number – whether positive or negative – is positive hence there is no square root of a negative number"*

The first known mathematician who dared to put such a mysterious number on paper to solve a similar riddle as that discussed above was the Italian mathematician **Gerolamo Cardano** (1501 – 1576)

Despite such widespread misrecognition; the square root of negative numbers continued to be popping up in mathematical expressions and in engineering applications.

One unexpected appearance of "i" is in Euler's formula:

$$e^{ix} = \cos(x) + i \sin(x)$$

The formula - which was discussed in chapter **6** - lead to the derivation of Euler's Identity stating that:

$$e^{i\pi} + 1 = 0$$

The identity sets a wonderful relation combining i with e, π and 1.

Another application is introducing $i = \sqrt{-1}$ as a factor to the "time interval" when calculating the distance separating two event in the relativistic four dimensional space-time continuum. This will be discussed in chapter 10.

It is always possible to relate any such square root of a negative number to the imaginary number i

For example:

$$\sqrt{-134} = \sqrt{134} * \sqrt{-1} = \sqrt{134}\, i$$

and

$$\sqrt{-100} = 10\sqrt{-1} = 10\, i$$

and so on, where $i = \sqrt{-1}$

Those numbers having *i* as a factor are called "**imaginary numbers**"

Imaginary numbers can be thought of a mirror reflection of real numbers because for real number **R** there is a corresponding imaginary number obtained by multiplying **R** into *i*

If i is raised to powers larger than 1 it will yield the following values:

$$i^2 = -1$$

$$i^3 = -i$$

$$i^4 = 1$$

$$i^5 = i$$

COMPLEX NUMBERS

Numbers comprising a mix of real and imaginary numbers in the form **a + b** *i* are called *Complex Numbers*. Examples for complex numbers are: **3 + 4** *i* and **x + y** *i*

While performing basic arithmetic operations on composite numbers you must sort out the real and imaginary parts as different. For example, to add two or more composite numbers; add the real parts in all numbers then add imaginary parts in complete separation. For example:

$$a + b\,i$$
$$+ \quad c + d\,i$$
$$+ \quad e + f\,i$$
$$+ \quad \underline{g + h\,i}$$
$$= \quad (a + c + e + g) + (b + d + f + h)\,i$$

Subtraction is performed in the same manner but with the sign of the composite number to be subtracted reversed.

To multiply (**a** + **b** i) into (**c** + **d** i) multiply each part in the first number by the two parts in the second one and substitute "– **b d**" for "**b d** i^2" so the product will be:

(**a** + **b** *i*) x (**c** + **d** *i*) = (**ac** – **bd**) + (**ad** + **bc**) *i*

To divide a complex number (**a** + **b** i) by a real number "**n**" you

should divide both real and imaginary parts by "**n**" so the result will

be: $\dfrac{a}{n} + \dfrac{b}{n} i$

How about dividing a complex number like (**a** + **b** i) by another complex number (**c** + **d** i) ?

To perform the division $\dfrac{a + b i}{c + d i}$ it will help that the imaginary part

in the denominator is diminished. This can be achieved by multiplying

both the numerator and the denominator by (**c** – **d** i) which is called

the conjugate of the denominator (**c** + **d** i)

$$\frac{a+b\,i}{c+d\,i} = \frac{a+b\,i}{c+d\,i} * \frac{c-d\,i}{c-d\,i} = \frac{(ac+bd)+(bc-ad)i}{c^2-d^2 i^2} = \left(\frac{ac+bd}{c^2+d^2}\right) + \left(\frac{bc-ad}{c^2+d^2}\right) i$$

Complex numbers are often represented in a two dimensional coordinate system, in which the horizontal axis indicates the real part of the number while the vertical axis indicates the imaginary part.

The complex number 8 + 6 i is represented as point "A" in the graph shown in figure 9-1.

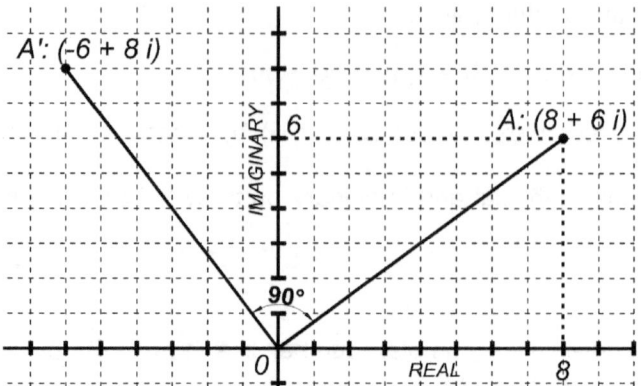

Figure 9-1: Graphical representation of complex numbers

Point "A" has the real and imaginary coordinates 8 and 6 at horizontal and vertical directions respectively.

The magnitude of a complex number is measured by the length of the line joining the point representing the number in the graph and the origin (point 0,0). This length is called the absolute value and it is always a real number.

The absolute value for the complex number **A = (8 + 6 i)** is represented using the symbol **|** in this manner:

| A | *(read as the absolute value of "A")* $= \sqrt{8^2 + 6^2}$ **= 5**

An interesting property of such coordinate representation is that if you rotate the line **OA** (from the origin to point **A**) **90°** anticlockwise direction, you will be mapping point **A** onto point **A'** which equals **A x** *i* :

A * *i* = **(8 + 6** *i***)** * *i* = **8** *i* + **6** *i*² = **(-6 + 8** *i***)** which is represented by point **A'**

CUBIC ROOTS OF "1"

The reader may object to this title questioning *"why there should be several cubic roots for "1"?*

The reason for to $x = \sqrt[3]{1.0}$ have several solutions (namely three) is that this equation is equivalent to:

$x^3 - 1.0 = 0$ which is a cubic equation hence there must be three roots (valid solutions) for x

To solve the above cubic equation, factor $(x^3 - 1.0)$ as follows:

$(x^3 - 1.0) = (x - 1) (x^2 + x + 1)$ The expression $x^2 + x + 1$ can be

factored using the general solution formula of quadratic equations:

$$X = \frac{-b \pm \sqrt{b^2 - 4ac}}{2a}$$ where a = 1, b = 1 and c = 1;

$$(x^3 - 1.0) = (x - 1) \left(x - \frac{-b + \sqrt{b^2 - 4ac}}{2a}\right) \left(x - \frac{-b - \sqrt{b^2 - 4ac}}{2a}\right) = 0$$

$$= (x - 1) \left(x - \frac{-1 + \sqrt{1^2 - 4}}{2}\right) \left(x - \frac{-1 - \sqrt{1^2 - 4}}{2}\right) = 0$$

hence: from which the cubic roots of **1.0** are:

- **1.0**

- $(\frac{-1}{2} + \frac{\sqrt{3}}{2} i)$

- $(\frac{-1}{2} - \frac{\sqrt{3}}{2} i)$

So there is one real root for the equation $x^3 - 1.0 = 0$ and a couple of conjugate imaginary roots. This is quite common. Imaginary roots always exist in conjugate couples – never single.

Furthermore, a couple of imaginary roots is always in conjugate form like **(a + b *i*)** and **(a – b *i*)**

Another interesting property in the two imaginary cubic roots of 1.0 is that if you square one of them you will get the other one:

$$(\frac{-1}{2} + \frac{\sqrt{3}}{2} i)^2 = (\frac{-1}{2} - \frac{\sqrt{3}}{2} i)$$

and:

$$(\frac{-1}{2} - \frac{\sqrt{3}}{2} i)^2 = (\frac{-1}{2} + \frac{\sqrt{3}}{2} i) ,$$

Either root is named ω which Is a the lower case of the Greek letter "omega" while the other one is named ω^2

Another interesting property is that the sum of the two imaginary roots = **-1** and the product of them is naturally ω^3 = 1, so you can have the three roots (ω, ω^2 **and 1**) present in a single formula in two different ways:

$\omega + \omega^2 + 1 = 0$ Also:

$\omega * \omega^2 = 1$

The Chapter Quiz

If $x = y^2$ and $y = x^2$,

$x \neq 0$, $y \neq 0$ and $x \neq y$

Find "**x**" and "**y**"

Discussion and solution of the Chapter Quiz

$x = y^2$ .... **(1)**

$y = x^2$ .... **(2)**

From (1) and (2):

$x = x^4$ .... **(3)**

Since $x \neq 0$ we can divide both sides of equation (3) by "**x**", hence:

$x^3 = 1$, from which $\quad x = \sqrt[3]{1.0}$

As discussed earlier; there are three cubic roots of 1.0 so which one of them we should consider?

If we take $x = 1.0$ we shall have $y = x^2 = 1$ also which contradicts with the given condition that $x \neq y$, hence we should pick another cubic root. Consider:

$$x = \omega = \left(\frac{-1}{2} + \frac{\sqrt{3}}{2}\, i\right)$$

and $y = x^2 = \omega^2 = \left(\frac{-1}{2} - \frac{\sqrt{3}}{2}\, i\right)$

Hint:

Another possible correct solution is to take:

$$x = \omega = \left(\frac{-1}{2} - \frac{\sqrt{3}}{2}\, i\right) \text{ and } y = x^2 = \omega^2 = \left(\frac{-1}{2} + \frac{\sqrt{3}}{2}\, i\right)$$

Chapter 10: Einstein's Relativity

Figure 10-1: Albert Einstein

SPACE AND TIME IN CLASSICAL PERSPECTIVE

According to the classical mechanics predominantly established by the great Isaac Newton; time is considered to be passing uniformly and externally through all objects in the universe and completely independent of the speed of these objects. Also according to classical mechanics, gravity between two objects is caused by an invisible attraction force between these objects proportional to the product of the masses of the two objects and inversely proportional to the square of the distance between the centers of gravity of these two objects.

Such a robust knowledge structure of classical mechanics has always been the main predominant reference for setting the rules of motion and gravity even after it had been challenged, generalized and fine-tuned by Albert Einstein (1879 – 1955) through his revolutionary papers on the Special Theory of Relativity (1905) and the General Theory of Relativity (1915).

In this context, it should be noted that in our normal life, Newton's classical mechanics describes the rules of motion with an adequate precision.

O+nly if an object reaches a high speed comparable with the speed of light, then the Newtonian algorithm will not work and relativistic approach should take place. This occurs in analyzing the motion of particles moving in very high speeds like – for example – in the case of sub-atomic centrifugal accelerators.

HIGHLIGHTS ON RELATIVITY

While the Special Theory addresses the special case where objects move in a uniform speed relative to one another; the General Theory generalizes the Newtonian mechanics and the Private Theory by addressing the cases of accelerating motion and explaining gravity in a rather revolutionary way.

Below is a brief highlights on basic elements of relativity as established by Einstein in his papers on the "Special Theory of Relativity" and the "General Theory of Relativity". More detailed description of the two theories will follow in the text.

- There is no experiment, device or methodology that can distinguish - from within an object's inertial frame - between the state of moving at a uniform speed and the state of being at rest

- The same laws of physics are observed by all the objects in the whole universe

- The speed of light in vacuum "c" is a universal constant fixed in all frames of references regardless of their relative speed or the speed of the source: c = **299,776** KM/sec = **186,300** Miles/sec

- Energy and mass are equivalent and are convertible to one another according to the equation $E = mc^2$, where E is the energy, **m** is the mass and **c** is the speed of light.

- Time is not an independent variable that flows uniformly throughout the universe. Time is the fourth dimension in a combined four dimensional space-time continuum

- Gravity is not caused by an invisible force between two bodies attracting one another as suggested by the Newtonian mechanics.

- The relativistic explanation of gravity is that the four dimensional space-time bends around large masses; which restricts the movement of other objects (of lesser mass) through paths of least resistance in that curved space

- There is no experiment that can distinguish between the effect of acceleration of a system and the effect of gravity. This is known as the Equivalence Principle

THE FOUR DIMENSIONAL SPACE-TIME CONTINUUM

We can imagine entities having two or three dimensions easily because we look at these entities from outside, being – ourselves - three dimensional beings. In fact the image we receive in our retina for a 3-D object is the projection of this object on our retina's 2-D surface. The brain takes the important role of making the 3-D interpretation intuitively. But leave imagination alone and let us try to visualize – that is to figure out the features and properties of – such a four dimensional object by stretching out our perception in an extrapolative way (*extrapolation of a relationship or graph is a technique often used by researchers to predict the features of successive incidences by extending the graph representing the predecessor ones*)

As an exercise of mind; we shall use extrapolation technique to visualize a simple object such as the 4D-cube and a 4-D space-time interval between two events.

Let us take a point in space "A" (that is a zero dimensional object), extend it to a two dimensional line AB that will be developed to a two dimensional square ABCD the continue by extending the square to the three dimensional cube by following the steps listed below and illustrated in figure 10-2:

a. Draw a point "B" at a distance "L". Join AB

b. Draw line segment DC parallel to line AB and at a distance L from it. Join AD and BC to have the square ABCD

c. Construct in the space a square EFGH to be congruent to square ABCD and at a distance L apart from it in a direction perpendicular to the plane of the square ABCD. Join lines AE, BF, CG and DH to construct the cube ABCDEFGH

To construct the four dimensional cube we shall start from the 3-D cube we just constructed and continue in light of the directions in step "c" above with sum adaptation to these directions. This step should now read as follows:

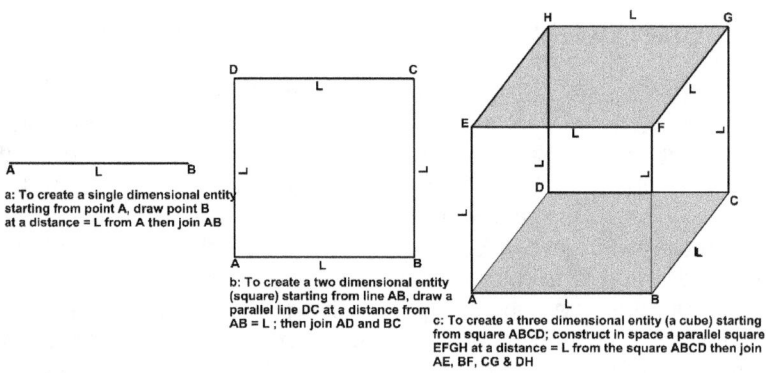

a: To create a single dimensional entity starting from point A, draw point B at a distance = L from A then join AB

b: To create a two dimensional entity (square) starting from line AB, draw a parallel line DC at a distance from AB = L ; then join AD and BC

c: To create a three dimensional entity (a cube) starting from square ABCD; construct in space a parallel square EFGH at a distance = L from the square ABCD then join AE, BF, CG & DH

Figure 10-2: The construction of a 3-D cube starting from a zero dimensional point

Construct -in the 4-dimensional space- another cube A'B'C'D'E'F'G'H' to be congruent to cube ABCDEFGH and at a distance L apart from it in a direction **perpendicular to the 3-D space of the cube ABCDEFGH**

Here we are facing a problem concerning the interpretation of the underlined statement above! We are familiar with a line drawn perpendicularly to a two dimensional plane; with all being constructed in a three dimensional space, but drawing a line perpendicularly to a 3-D space is something we cannot imagine because we have never seen. However, we should agree that the statement makes sense.

To convert the two dimensional square to a three dimensional cube, we had to construct another square congruent to the original one and apart from it *in a direction perpendicular to its plane*. Indeed, we know that in the 3D space there is a direction Z which is perpendicular to plane XY, so why to dismiss the analogy in the four dimensional space. Why can't we have – in the 4D space - a direction T which is mutually perpendicular to each of the orthogonal directions X, Y and Z ? For such possibility to defy the common sense or to be unimaginable is not a good reason for us to dismiss it altogether.

Despite the complete lack of familiarity with such 4-D monsters and even the ability to imagine them; we can still explore their structures and visualize their properties

A schematic drawing for a four dimensional cube is shown in Figure 10-3 constructed in the way proposed above; that is by copying the 3-D cube ABCDEFGH to a position displaced a distance "L" and joining each couple of their corresponding vertices by straight lines (dashed).

It will not be difficult for us to find that there are 16 vertices and 32 edges in that monster four dimensional cube.

Let us stretch out our perception to grasp the features of a four dimensional sphere.

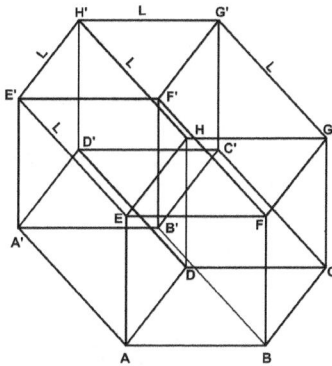

**Figure 10-3:   A
hypothetic image for a
4-D cube**

Imagine that you are projecting a three dimensional crystal globe on a two dimensional screen using parallel light bundle.

You will get a two dimensional picture in which both images of light facing and light remote hemispheres are merged as shown in figure 10-4. As seen in the figure, such merging has resulted in having the

Australian continent appearing closely between Europe and North America (yet in a different color due to light refraction through the crystal globe)

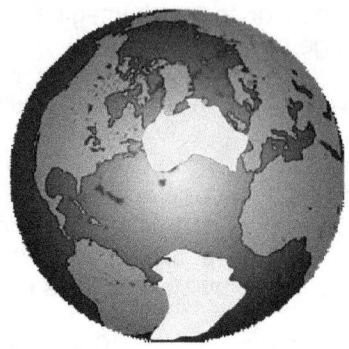

Figure 10-4:

**2-D projection for
a 3-D crystal
globe**

Extrapolating from the projection of 3-D sphere to the projection of 4-D sphere we can deduce that the latter projection will result in two merged three dimensional spheres having their two outer surfaces coinciding completely with each other.

We shall pursue the same extrapolation technique -followed in creating the four dimensional cube and sphere - to grasp the meaning of a *space-time interval* between two events in the four dimensional space-time continuum. In doing so it would be convenient to start with the simple task of finding the distance between points "a" and "b" located in a familiar two dimensional plane.

Points a and b in figure 10-5 have the X, Y coordinates (Xa, Ya) for "a" and (Xb, Yb) for "b"

The projected length of line segment ab on the X and Y axes respectively are (Xa – Xb) and (Yb – Ya) respectively. The line segment ab is obtained by applying the Pythagorean Theorem as follows:

(ab)^2 = (Xa – Xb)^2 + (Ya –Yb)^2 from which

$$ab = \sqrt{(Xa - Xb)^2 + (Ya - Yb)^2}$$

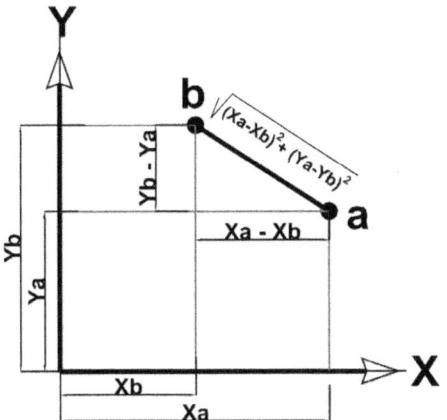

Figure 10-5:

Distance

separating two

points in a two

dimensional

plane

Moving forward to measuring the spatial distance between two points a and b located in a 3-D space as shown in figure 10-6, I would refer the reader to the case of determining the hypotenuse D of the cuboid shown in figure 1-13 (Chapter One). From the examples represented by figures 1-13 and 10-5 and considering the line segment in the three dimensional system of coordinates shown in figure 10-6, following can be deduced:

$$(ab)^2 \quad = (Xa - Xb)^2 + (Ya - Yb)^2 + (Za - Zb)^2 \text{ hence}$$

$$ab = \sqrt{(Xa - Xb)^2 + (Ya - Yb)^2 + (Za - Zb)^2}$$

|Xa – Xb|, |Ya – Yb| and **|Za – Zb|** are the projections of the line segment "**ab**" on the three orthogonal mutually perpendicular planes **XY**, **YX** and **YZ** respectively. **|Xa – Xb|** means the absolute value of **(Xa – Xb)** that results from subtracting the smaller of Xa and Xb from the larger, which should always be positive.

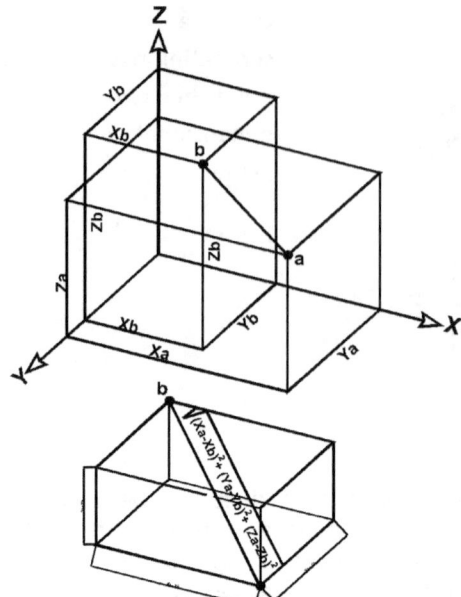

Figure 10-6: Distance separating two points in a 3-D space

We can move now to the bold step of finding the interval "**a b**" in Einstein's unified four dimensional space-time continuum.

It is not unusual for us to relate a certain event to three spatial dimensions and a forth temporal one. If you want to describe the position of an event – say breakout of a fire in a residential apartment - you often mention the time it happened along with the three spatial coordinates.

If event "**a**" is the breakout of fire, you say in your description: *"the fire took place at a building **500** meters to the North, **150** meter to the East, at the **5**th floor at **5:36:32** PM"*.

Notice the three spatial dimensions mentioned here:

Xa: *500 meters to the North*
Ya: *150 meters to the East,* and **Za**: *at the 5th floor (say 20 meter above sea level),*
The time dimension is also described in: **ta**: *at 5:36:32 PM*

Assume that event "**b**" is a lightning strike that hit a telecommunication tower built on high mount at the following space-time coordinates

Xb: *6500 meters to the North*
Yb: *3550 meters to the East,* and **Zb**: *200 meters*
The time dimension is also described in: **tb**: *at 5:36:30 PM*

Points "**a**" and "**b**" in our four dimensional space-time continuum shall therefore represent two events; each of which is described using the three spatial dimensions (X, Y and Z) in addition to the fourth temporal dimension **t**.

You may raise an objection here: "*how can we use metric units to specify the three spatial coordinates and temporal units such as hours, minutes or seconds to specify the time dimension?*" The objection is undoubtedly legitimate!
This technicality has been resolved by expressing the time "t" in terms of the distance travelled by light during such time.

For example, if the time interval is 10 seconds, you express it as the distance travelled by light during this 10 seconds, which is 10 x **299,776 = 2,997,760** km (or 10 x **186,300 = 1,863,000** miles).

Converting time units such as seconds to distance units such as kilometers is simply done by multiplying the time (in seconds) by the speed of light c = **299,776** KM/sec = **186,300** miles/sec

The four dimensions can therefore be expressed in the same distance measuring units (meter, kilometer, foot or mile).

Light speed has been chosen as a conversion reference because it is a universal constant.

You may still object to that seemingly irrational mathematics which obliterates the physical differences between space and time as such. After all, we really need to stretch our perceptions far to visualize the world of four dimensions which we have never viewed – and will never be able to view- from outside and can only perceive its properties by analyzing the physical phenomena underlying its four dimensions.

Hermann Minkowski (1864 – 1909) realized in 1907 that the Special Theory of Relativity introduced two years earlier by his former student Albert Einstein could best be presented in a four dimensional space-time continuum, in which time is not as independent of space as used to be presented by the Newtonian theories. Later in 1915 Einstein published his paper on the General Theory of Relativity which explained the properties of such four dimensional space-time continuum, its curvature around big masses and gave profoundly unconventional explanation of gravity.

It was also Minkowski who proposed that the time dimension – in the 4D space time geometry – should be multiplied by $i = \sqrt{-1}$

such that it would be distinguished from the other three spatial dimensions (I would refer the reader to Chapter 8 in which imaginary and complex numbers were introduced)

On basis of foregoing discussions, the two events "a" and "b" in figure 10-6 are separated by a space-time segment ab where:

$$\mathbf{ab} = \sqrt{(\mathbf{Xa} - \mathbf{Xb})^2 + (\mathbf{Ya} - \mathbf{Yb})^2 + (\mathbf{Za} - \mathbf{Zb})^2 + (i\mathbf{Ta} - i\mathbf{Tb})^2}$$

and since $i^2 = \left(\sqrt{-1}\right)^2 = \text{-1}$, above equation may take the form:

$$\mathbf{ab} = \sqrt{(\mathbf{Xa} - \mathbf{Xb})^2 + (\mathbf{Ya} - \mathbf{Yb})^2 + (\mathbf{Za} - \mathbf{Zb})^2 - (\mathbf{Ta} - \mathbf{Tb})^2}$$

Where **(Ta – Tb)** is the time interval between events **a** and **b** converted to linear space units by multiplying the time interval in seconds **(ta – tb)** by **c = 299,776 KM/sec**

The three spatial intervals in kilometer are:

$$\begin{aligned}
\mathbf{Xa - Xb} &= \mathbf{(0.500 - 6.500)} &&= \mathbf{-6.000\ km} \\
\mathbf{Ya - Yb} &= \mathbf{(0.150 - 3.550)} &&= \mathbf{-3.400\ km} \\
\mathbf{Za - Zb} &= \mathbf{(0.020 - 0.200)} &&= \mathbf{-0.180\ km}
\end{aligned}$$

Time interval **(ta – tb) = 2** seconds to be converted to metric units by multiplication into **c = 299,776,000 meter/sec**, hence the converted time interval:

Ta – Tb = 2*299,776   = 599,552   km

The space-time interval (ab)

$$= \sqrt{(Xa - Xb)^2 + (Ya - Yb)^2 + (Za - Zb)^2 - (Ta - Tb)^2}$$

$$= \sqrt{(6.00)^2 + (3.40)^2 + (0.180)^2 - (599,552)^2}$$

$$= \sqrt{47.6 - 359462600704} \quad = 599552\sqrt{-1} = 599552\,i$$

You will notice that the result of the calculations shows an imaginary value for the (**a b**) interval. This is due to the two events taking place in our planet; that is at a relatively a very short spatial distance. Remember that the theories of relativity address much broader cosmic spaces and events spatially separated by several light years.

Consider another example in which event "**a**" is the collision of a meteor on Mars on 14th April 2015 at **6:54** PM as detected in an Observation Center at an island located somewhere in the Pacific ocean. The distance from the meteor to the observation center when the collision happened was **384,000,000 km**. Event "**b**" was a car accident that happened in that same observation center at 7:02 PM i.e. only **8** minutes after the collision.

The spatial distance between events a and b,

$$\mathbf{Ds} = \sqrt{(Xa - Xb)^2 + (Ya - Yb)^2 + (Za - Zb)^2}$$

$$= 384,000,000 \quad \text{km}$$

Converted time interval $\mathbf{Dt = Ta - Tb} = 8*60*299,776$

$$= 143,892,480 \quad \text{km}$$

The space-time interval (ab)

$$= \sqrt{(384,000,000)^2 + (143,892,480\,i)^2} = 356,021,002 \text{ km}$$

THE SHOCKING RESULTS OF MICHELSON-MORLEY EXPERIMENT

The experiment performed in 1887 by Albert Michelson and Edward Morley was intended to detect a variation in the light relative velocity as it moves forward and backward in the direction of light carrying ether as opposed to its relative velocity as it moves perpendicularly to it.

The 19th century physicists believed that light waves are carried in space by the so called "Light Carrying Ether" in a similar way voice waves are carried by air or solids. Since sound cannot be transmitted in vacuum and needs a tangible medium to carry it through its destinations, there must be a tangible medium in space – so they believed - to carry light through its destination, and the ether was that hypothetic medium.

Ether was assumed to be filling up every space in the universe. It was assumed to behave like a vibrating solid to transmit the light waves yet to have the properties of a fluid liquid that lacks any resistance to celestial objects moving through it. Among these celestial objects is our Earth which moves in its orbit around the sun through ether at a speed of 30 km/sec. It was accordingly assumed that the ether wind blows on earth in a similar way as wind blows on you when you ride onto a speeding open carriage in a windless day, and as symbolically illustrated in

The experiment – shown schematically in figure 10-8 – was setup on a granite block floating on a pool of mercury.

Earth movement

@ 30 km / sec

Ether wind

Figure 10-7: Ether wind blows in a direction opposite to movement of earth

THE EXPERIMENT

A light beam is emitted from a source of light at "**a**" towards a semi-transparent mirror at "**c**" -that reflects nearly 50% of light emitted to it- placed in the center of the circular granite worktop. The portion of light reflected by the semi-transparent mirror "**c**" – let us call it portion "**R**" - goes to mirror "**d**" at the periphery and reflected back to "**c**" which will partly allow it to continue through to a telescope fixed at point "**f**". This portion of light will be travelling across the ether wind in its trip from "**c**" to "**d**" and in the return trip from "d" to "**c**".

Remaining 50% portion of the light emitted from "**a**" and passing through "**c**" - let us call it portion "**P**" - will reach a mirror "**b**" at periphery which bounce it back to "**c**" that will partly reflect it to the telescope at "**f**".

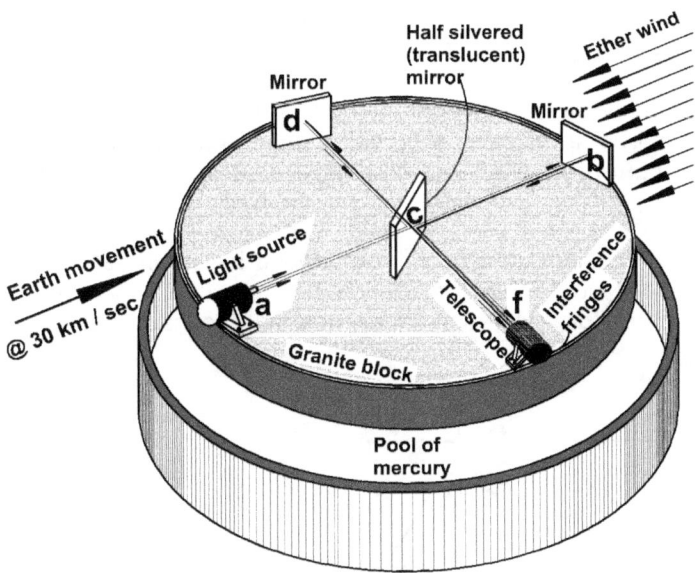

Figure 10-8: Michelson - Morley experiment

This portion of light will be travelling against the ether wind in its trip from "**c**" to "**b**" and along with it in its return trip from "**b**" to "**c**"

It is important to remember that this whole experiment is strictly concerned with comparing the delay caused to portion "**R**" of the light for its travelling across the ether wind in its return trip from **c** to **d** and from **d** to **c**; with the delay caused to portion "**P**" of the light for its travelling in the opposite direction of the ether wind in its trip from **c** to **b** then along with it in its return trip from **b** to **c**. That is why the trips from **a** to **c** and from **c** to **f** have not been considered because they are common for the two portions of light hence equal delay – if any – will be caused to them in these segments. It should also be noted that **cb = cd**

According to optical theories, the two portions of light beams R and P will interfere and will form an array of dark and light fringes that can be seen by the telescope placed at point "f".

If the two portions return (from d & b) to the translucent mirror at "c" simultaneously; the bright fringe will be in the center of the image.

Otherwise, if one of portions R and P is delayed with respect to the other the fringes will be shifted off the center.

To assess the delay caused to portion **R** travelling across the ether wind, consider a boat at a speed V crossing a river having a breadth H and stream current of a speed **v** as illustrated schematically in figure 10-9

The increase in crossing time "t_o" will be caused by the increase of the path due to the boat moving side-way in a direction opposite to the current in order for it to reach to the desired destination across the river.

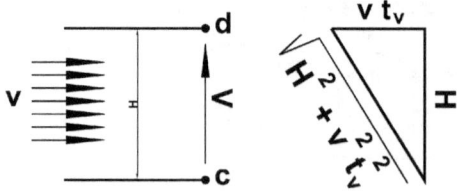

Figure 10-9: Effect of stream current in delaying a boat crossing the river

Assume that **H** is the width of the river, **V** is the boat speed, **v** is the velocity of the current across which the boat moves, t_o is the crossing time had not been there any current and t_v is the time the boat will take to cross the river across the currents.

Slanted distance to be travelled by the boat = $\sqrt{H^2 + v^2t^2}$

$$t_v = \frac{\sqrt{H^2+v^2tv^2}}{V} ,$$

$$(t_v)^2 = \frac{H^2+v^2tv^2}{V^2}$$

$$t_o = H/v = t_v\sqrt{(1 - \frac{v^2}{V^2}}$$

$$t_v / t_o = \frac{1}{\sqrt{(1-\frac{v^2}{V^2}}}$$

Returning to the Michelson-Morley experiment we can apply the same formula to find the delay ratio caused by the ether wind for portion "R" of light to travel across it in a round trip from c to d; by substituting T_R (time taken by light to travel across ether wind) for t_v , T_o (time taken by light to travel in absence of ether wind) for t_o and c (light speed) for **V**

$$T_R / T_o = \frac{1}{\sqrt{1-\frac{v^2}{c^2}}} \quad \text{..} \quad (1)$$

Portion "P" of light travels from c to b against ether wind presumably at a relative speed of c − v (where v is the speed of ether wind), and returns to c at a relative speed of c + v (we will see later how erroneous were these assumptions). Time taken by light in the round trip is T_P while time supposed to be taken in the trip had not ether wind been there is T_o

$$T_0 \quad = \frac{2H}{c}$$

$$T_P \quad = \frac{H}{c+v} + \frac{H}{c-v} = \frac{2cH}{c^2 - v^2}$$

$$T_P / T_o = \frac{1}{\left(1-\frac{v^2}{c^2}\right)} \quad\quad \dots\dots\dots\dots\dots\dots\dots\dots\dots\dots\dots\dots\dots\dots \text{(2)}$$

Now we can find the relative delay between portion **R** of the light and portion **P** by dividing the right hand expression of equation (1) by the right hand expression of equation (2):

$$T_R / T_p = \frac{1}{\sqrt{1-\frac{v^2}{c^2}}} \quad\quad \dots\dots\dots\dots\dots\dots\dots\dots\dots\dots\dots\dots \text{(3)}$$

Such expected delay between the light beams of portions R and P – no matter how trivial it is - would have caused the interference fringes in the monitoring telescope to be shifted.

However, the **experiment did not show any shift in these fringes, hence no impact of the ether wind was confirmed**. The same result persistently appeared in subsequent repetitions of the experiment.

The surprise was utterly shocking not only to Michelson but to the 19th century physicists in general because it constituted a blow to the concept of relativity of the speed of light.

Everyone was sure about the precision of the experiment but no one could explain why the light beam portion **P** had not been delayed beyond the light beam portion **R** as expected.

A bold explanation was made then that the thick marble worktop must have been shrunk in the direction of the earth movement (a b) for a factor of its original length equaling $\sqrt{1-\frac{v^2}{c^2}}$ (see equation (3) above) hence the delay was totally eliminated by such shortening.

Shortening of an object in the direction of its motion as such is known as **Lorenz-FitzGerald** contraction which attributes the contraction to purely mechanical factors.

However, it turned out subsequently that the assumption was correct as far as the shortening initiator, direction and magnitude are concerned. Shortening of objects in the direction of movement was subsequently supported by Einstein in the Special Theory of Relativity but not on basis of mechanical strain as was assumed earlier.

Einstein's theory establishes that the space holding an objects which is in motion relative to an observer; contracts in the direction of movement to a length factor equaling **Lorenz** contraction factor.

It should be made clear that this contraction is detected only by an observer at rest relative to the moving object.

For example if a spaceship carrying an astronaut is moving at a high speed – say 87% of the speed of light relative to a ground observer,

$$\textbf{Lorenz factor} = \sqrt{1-\frac{v^2}{c^2}} = \sqrt{1-0.87^2} \simeq 0.5$$

The astronaut will not feel any shortening in the dimensions of the spaceship or the objects onboard including himself, but a ground observer will notice that the spaceship has shrunk to half its length in the direction of movement.

THE SPECIAL THEORY OF RELATIVITY

A paper on the Special Theory of Relativity was published by Albert Einstein in the year 1905. To grasp the significant change brought about by the theory let us first have a short primer on the classical concept of relativity.

If you sit in train that moves at a uniform speed of **50** meter/second, you will be seen by an observer standing beside the track outside as moving at that same speed. If you decide to run inside the moving train at a speed of **10** meter/second in the direction of its movement the same observer will see you moving at a speed of **50 + 10 = 60** meter/second.

But if you decide to run at the same speed but in the other direction the your movement relative to this same observer will be **50 – 10 = 40** meter/second. You are still in that train and feel bored so you decide to sit in one of the train's seats, to open a window beside you and -for some reason- to throw an object – say an apple – from the window at a speed of **5** meter/second in the same direction of the train's movement.

Another observer outside the train – the previous one is out of sight by now - will notice that the apple moves at a speed of **50 + 5 = 55** meter/second.

Eventually you realize that throwing apples is not amusing enough so pull your revolver from its grab and decide to shoot a bird which is sitting on a tree which is located away in the direction of movement. You know that the speed of the bullet is **200** meter/second; however a third observer standing outside the train will find that the bullet is rapidly moving at a relative (to him) speed of **50 + 200 = 250** meter/second.

At this moment you will feel that bullet shooting is not amusing enough so you will pull out your laser gun (which you always keep with in travels) and target another bird on another tree.

You know that the speed of a laser beam is the same as the speed of light which is nearly 300,000,000 (three hundred million) meters/second. Can you estimate the speed of the laser emitted from your gun; as detected by a fourth observer standing outside the train?

Following foregoing examples you may tend to assume it as 50 (the speed of the train) plus 300,000,000 (the speed of light) = 300,000,050 meter/second.

Here Einstein's soul will show a strong objection, for the speed of light is not relative as in the case of all other speeds. The speed of light in vacuum "c" is a universal constant fixed in all frames of references regardless of their relative speed or the speed of the source.

In the example just given, all the objects inside a train moving at **50** meter/second will acquire the same speed initially and if any of these objects moves relative to the train and in its movement direction at a speed "**s**"; its speed relative to an outside standing observer shall be **50 + s**.

You can say that the speeding train gives an advance push – initial speed – to all the objects onboard. The laser gun emitting light is the only exception to that because no advance push can ever be given to the light. This is a prominent principle upon which the Special Relativity is based.

Another important principle of the theory establishes that the same laws of physics are observed by all the objects in the whole universe

The theory of special relativity is concerned with objects moving in straight lines and at uniform speeds. You may ask: "*What about objects at risk?*" The answer from a relativistic perspective is that there is no experiment, device or methodology that can distinguish - from within an object between the state of moving at a uniform speed and the state of being at rest

The example of the moving train describes hypothetical virtual actions or events. You will perform the experiments discussed in the example only in your thoughts!

"**Thought Experiments**" is a known technique used in solving mathematical problems and in proving conjectures.

The technique was favored by Einstein and was used by him to validate basic relativistic hypothesis as will be discussed shortly.

It was also used in Chapter Three of this book to find the shortest path of line segments connecting a number of points by linking the points by strings made taut by the same force.

THE LIGHT CLOCK THOUGHT EXPERIMENT

The light clock thought experiment is a system that comprises a source of light pulses attached to a clock, a mirror, another clocks and observer(s) to measure the time taken for a light pulse emitted from a source in the ceiling of a train to reach the floor where it is bounced by a mirror then to return back to another clock fixed at the ceiling. Figure 10-10 schematically illustrates the functioning of a light clock fixed in a train moving at speed "**v**" in the two cases:

a. The observer is inside the train
b. The observer is outside the train

The experiment is designed to demonstrate the phenomenon called time dilation.

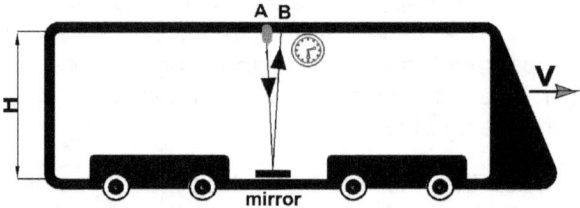

a. The observer is inside the train

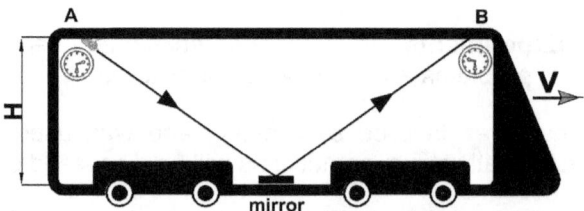

b. The observer is outside the train

Figure 10-10: The light clock Thought Experiment

An observer inside the train will detect the light pulse travelling vertically from the source to the mirror (distance H) and from the mirror back to the ceiling (another distance H). The time taken by light in the round trip:

$$To = 2H/c, \quad (To)^2 = 4H^2 / c^2 \quad \text{............................ (1)}$$

The observer outside the train (lower part of figure 10-10) will figure that light must be taking the longer slanted path from point A to the mirror then from the mirror to point B. This is due to the observer being in a different reference frame relative to which the train is moving and light has to take this path to meet the now moving mirror then the now moving point B (to the inside observer in the upper part of figure 10-10; neither the mirror nor point B are moving)

Applying the Pythagorean theorem, we can find that the length of the

slanted path is $\qquad \sqrt{4H^2 + (v.Tv)^2}$

where v is the speed of the train, H is its internal height and Tv is the

time taken by light to travel along the V-shaped slanted path. Tv can

also be obtained by dividing this slanted path by the speed of the

light "c"

$$Tv = \frac{\sqrt{4H^2 + (v.Tv)^2}}{c} \quad \text{from which: } (Tv)^2 = \frac{4H^2 + v^2(Tv)^2}{c^2}$$

Hence; $(Tv)^2 = \dfrac{4H^2}{c^2 - v^2}$ $\qquad$ .... (2)

From equation (1) and (2)

$$\frac{To}{Tv} = \sqrt{1 - \frac{v^2}{c^2}} \qquad \text{............. (3)}$$

As you see; **Lorenz** contraction factor emerges again here.

The results of the Light Clock Thought Experiments are quite significant.

An observer outside the train with notice that light emitted by a source at the ceiling of speeding train will bounce on a mirror on its floor and hit the ceiling again in a V-shaped diagonal lines longer the light path detected in the same train wagon by an inside observer. Since the light speed is a universal constant we have to accept that the time elapsed from emission to return as counted by an external observer is longer than the time elapsed between the two events as counted by an observer onboard of the train. The problem typically encountered in perceiving the theory of relativity is that it defies the common sense as has been deeply established in our minds. None of us has ever been travelling at a speed close to that of the light hence it is deeply perceived that time flows uniformly in complete independence of movement.

A time interval between two events is differently measured by observers at different – relatively moving – frames of reference. Let the two events be the emission of a light pulse from a source at point "A" at the ceiling of the train (figure 10-10) and the return of bounced light to point "B" at the same ceiling. This time interval appears longer for an external observer than it does for an internal one. The internal observer will not notice any change due to over-speeding of the train (if he doesn't look through a window he may not even know whether the train is moving).

On the other hand; the external observer will notice that time passes slower inside the train. Clocks inside the train will be seen as moving slower. Even biological functions such as heart beat and respiration of the train passengers – as detected by an external observer - will be slower.

For an external observer, time in the speeding train will generally flow at a slower pace equaling Lorenz contraction factor.   This phenomenon at which time flows at different paces in different inertial frames is known as "Time Dilation".

It is a fully symmetrical phenomenon.

Passengers of the train have the right to consider themselves at rest while everything else around them is moving underneath their train. Accordingly, they will notice that time passes slower in that speeding universe outside the train.

In our day-to-day life we are used to perceive time as an independent constant that flows at the same pace everywhere; because the speeds we experience are infinitesimal if compared to the speed of light hence the relativity of time concept seems to violate common sense.

To visualize the impact of time dilation at inertial frames moving at a high speed relative to an observer; imagine that you decide to make a trip to a remote planet at another galaxy using a space rocket that can travel at a very high speed = **0.9999999** of the speed of light.

Knowing that the distance from earth to that planet is **10** light years, you will expect that the round trip aboard a space rocket moving – almost – at the speed of light (*the word "almost" here is very important because moving at the speed of light exactly is impossible*) will take you **20** years.

With such an incredible speed the **Lorenz** contraction factor will be

$$\sqrt{1 - \frac{v^2}{c^2}} = \sqrt{1 - 0.99999999^2} = 0.000141421 = 1/7071$$

So the time in your space rocket will be slowed down by a factor of **7071**.

While you will not feel any change, observers in a control center on earth will detect that the rocket clock, your wrist watch, your heart beat, your respiration and the speed of all your biological and mental processes will slow down at the rate of **1: 7071**. From the viewpoint of earth observers the trip will take 20 years, but you will swear that it took you only 25 hours. You will only have three meals and will sleep once during this 25 hours you have spent in the round trip to that remote planet. However, when you arrive back to earth - feeling that the trip started from earth the day before - you will find that your friends and relatives have been spending 20 years waiting for you with aging signs appearing on them.

RELATIVITY OF SIMULTANEITY THOUGHT EXPERIMENT

If two events are noticed by an observer to be occurring simultaneously; they might not be seen so by an observer at another frame of reference hence simultaneity is relative.   In the Relativity of Simultaneity Thought Experiment – illustrated in figure 10-11 – a light pulse is emitted from a source in the center point of a wagon in a train speeding uniformly.

As the light – travelling at both directions along the wagon – reaches either end of the wagon a door at that end will open automatically.

An observer inside the wagon (left side - "a' of the figure) who is at rest relative to the train will notice that the doors will open simultaneously at the two ends of the wagon. This is quite normal to him because he sits in the center point at an equal distance to the two ends.

An observer outside the train (right side - "b' of the figure) who would be coincidently standing opposite to the light source as the train passes by him; will see the back door opening first because the train is speeding forward towards the light hence the light will hit the back door first (the second of the three images in group b). Shortly after this happens the external observer will see the front door opening when the light pulse will reach it.

The conclusion is that the two events of opening the back and front doors of a speeding train wagon will be seen by a passenger sitting at the center point as simultaneous and will not be seen so by an external observer who will insist that back the door opened first. The experiment manifests the Relativity of Simultaneity.

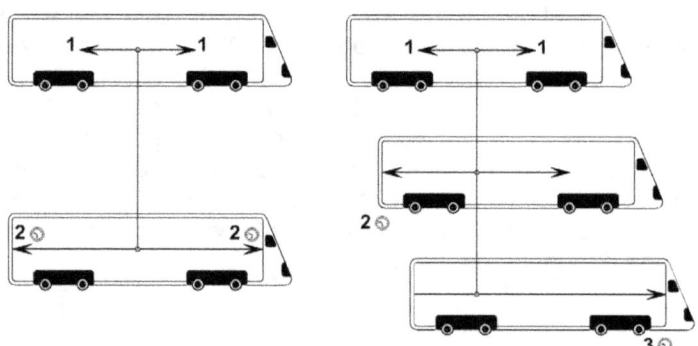

a. The observer is inside the train   b. The observer is outside the train

Figure 10-11: The relativity of simultaneity Thought Experiment

VELOCITY, MASS AND ENERGY

Mass is the amount of matter in an object. There are two methods to measure the mass of an object.

First is to weigh it as we always do in our day-to-day life. While the domain of applying the Theory of Relativity is the whole universe, such a method – based on weighing the object - is not convenient for application universally. That is because an object will weigh on the moon about eighth its weight on earth and much less in the deep space outside appreciable fields of gravitation. The second method to measure mass is based on Newton's Second Law of motion stating that a force "**F**" exited on an object having mass = "**m**" will cause it to move in acceleration "**a**" where **F = (m . a)** hence: mass "**m**" = F/ a.

The mass – according to this method – is the amount of force needed for the object to reach to the unit acceleration. For example, an object having a mass of **20** kg needs to be pushed (or pulled) by a net force (in addition to that required to counteract any resistance) of **20** Newton such that it would be moving at an acceleration of **1 meter/sec^2**

Mass measured by this method is called "inertial mass" as opposed to "gravitational mass" obtained by weighing. This second method – based on inertial mass - gives the same result regardless of the location of the object and whether it is on earth, on moon or elsewhere.

According to the Special Relativity, the mass of a moving object as detected by a stationary observer - will be increased in an inverse proportion to the Lorenz ratio:

$$\sqrt{1 - \frac{v^2}{c^2}}$$

For example, if the relative speed between the moving object and the observer = **0.87 c** (**c** being the speed of light), the mass of the moving object will increase at the ratio:

$$\frac{1}{\sqrt{1 - \frac{v^2}{c^2}}} \approx \frac{1}{\sqrt{1 - .87^2}} = 2$$

This means that the mass of the moving object (as measured by an external observer) will be doubled.

If relative speed between the moving object and the observer increases further mass in the moving object will continue to increase drastically, so the energy required for it to maintain the same speed will also be increased proportionately. When relative speed approaches c (the speed of light) mass will approach

$$\frac{1}{\sqrt{1-1}} = \infty \text{ (infinity)}$$

That is why the speed of light is the absolute maximum speed that cannot be exceeded by any object or even any other phenomenon.

According to the classical mechanics; if two objects **A** and **B** are moving at speeds v_a and v_b at opposite directions towards each other, we would calculate the relative speed between the two objects (v_{ab}) as: $v_{ab} = v_a + v_b$ in what is known as the rule of adding speeds.

If a space rocket "**A**" moves from east to west at a speed of **0.75 c** – relative to an observer in an Earth base and if another space rocket "**B**" is moving at the same time from west to east at a speed of **0.75 c** relative to the same observer; the Earth bound observer may assume that the two space rockets move towards each other at a relative speed = **0.75 c + 0.75 c = 1.5 c**. That is **1.5** the speed of light according to the rule of adding speeds. This is an obvious violation to the principles of the Special Relativity that set the speed of light as an absolute barrier.

The Special Theory of relativity has generalized the rule of adding speed to suit high speeds approaching light in the following way:

Take as an example the case of object **A** moving at speed **Va** = ¾ **c** (where c is the speed of light) and object **B** moving at the same speed but from the opposite direction X.

According to classical dynamics; relative speed between one another would have been obtained by simple addition in which case relative speed **Vab** would have been equal to ¾ **c** + ¾ **c** = **1.5 c**. That is one and half the speed of light.

Such speed is obviously rejected from relativistic perspective for its breaking the unreachable barrier of the speed of light.

According to the principles of special relativity, relative speed between the two objects of this example is:

$$\textbf{Vab} \quad = \frac{Va + Vb}{1 + \frac{Va.Vb}{c^2}}$$

$$= \frac{3/4 + 3/4}{1 + 9/16} \; c = \textbf{0.96 c}$$

That is still below the unbeatable threshold of **c**

Notice that in case speeds **Va** and **Vb** are very small compared to **c** as in most of the speeds on earth, $Va.Vb/c^2$ can be neglected which would bring the rule of adding speeds to its classical form: **Vab = Va + Vb**

An important result of Special Relativity is the concept of unification between mass and energy. Under certain conditions; mass can be converted to energy and under other certain conditions energy can also be converted to mass. Pre- relativity physicists used to believe that the total amount of mass in the universe is constant in what is known as the law of "*mass conservation*". They also believed that the total amount of energy is also constant in what is known as the law of "*energy conservation*". These two laws are now substituted by the law of "*mass and energy conservation*".

When heat energy is put into a teapot; water molecules will acquire kinetic (movement) energy that makes them move rapidly in an apparent disorder. A part of this energy will be converted to mass and the result is a slight increase in the mass of the water.

If you wind the spring of your watch, a part of the energy you have added as such will be converted to mass so the mass of your watch will be infinitesimally increased. This will obviously be unnoticed and will be reversed when the potential energy added by winding will be used out.

The nuclear reaction that takes place in the sun and in the explosion of a hydrogen bomb constitutes a case in which mass is transferred to energy. In such reactions; two hydrogen atoms are fused to create an atom of Helium associated with the release of energy.

The equation that represents such mutual conversions between mass and energy is the most famous: $e = m c^2$

e being the energy, **m** is the mass and **c** is the speed of light.

Einstein used a very smart thought experiment to derive this equation. He considered a particle of light called in the terminology of physics the "photon". It was established earlier by James Clerk Maxwell (1831 – 1879) that a photon has no mass but has a momentum. However, we know that **Momentum = Mass x Speed,** so how can we have a non-zero product of two factors one of them equals zero? Einstein reconciled the discrepancy by assuming a virtual mass "**m**" for the photon equivalent to the energy it has.

The thought experiment runs as follows: Imagine a box of mass "**M**" and inside length "**L**" is floating in the deep space away from any magnetic field.

A photon is emitted from a point on the left end of the box. The momentum of the photon P_{photon} equals the energy E divided by the speed of the photon (which is the speed of light) c:

$$P_{photon} = E/c$$

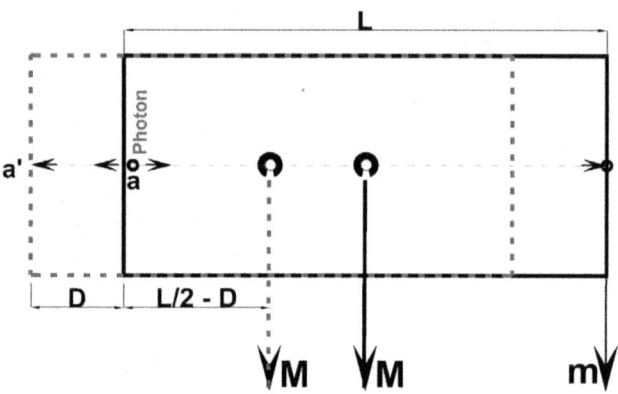

Figure 10-12: The Thought Experiment used to derive the formula e = m.c^2

As a reaction for the emission of the photon; the box will set back at speed "**v**" so its momentum: P_{box} **= M. v**

Time taken for the photon to travel to the right side of the box:

t_o **= L/c** during which the box would have moved in the left side direction a distance "**D = v . t_o**" hence **v = D/ t_o**

By conservation of momentum; $P_{photon} = P_{box}$ hence

$E/c = M. v = M. D/ t_o$, and since $t_o = L/c$ therefore:

$E/c = (M. D. c) / L$, hence

$MD = EL / c^2$

The total mass in the system = M + m, and since the center of mass at the startup of the experiment should remain at the same position at its end, moments of masses about any point at startup should be the same at the end. Taking the moment of masses @ point "a" in figure 10-12:

$ML/2 = m.L + M (L/2 – D)$, hence $D = m.L /M$ Replacing D by m.L /M in the above equation we get:

$mL = EL/c^2$ hence $E = mc^2$

SPECIAL RELATIVITY WRAPPED UP

Assume that two racing cars are approaching each other at the incredibly high speed of 260,000 km / second (relative to each other) as shown in the picture in figure 11-13.

The pictures depicts the image as seen by the driver of the car to the right hand side who looks frightened for seeing the other car so distorted (It was drawn by him shortly after he has recovered from the shock)

To visualize the impact of driving at this high speed relative to the other car, we would find the

$$\text{Lorenz factor} = \sqrt{1 - \frac{v^2}{c^2}} = \sqrt{1 - \frac{260,000^2}{300,000^2}} \simeq 0.5 \text{ in this case}$$

Figure 10-13: Two cars moving at a relative speed of 260,000 km/sec as seen by the driver at right hand side

A driver in either car will not sense any change happening within his inertial frame (the car he is driving).

However each one will report the following changes in the other car as detected by him:

- It has contracted to half as long

- Its clock is running half as fast and its seconds are twice as long

- The driver moves like characters in a slow motion film

- Its mass is twice as large

THE GENERAL THEORY OF RELATIVITY

Two years after publishing the Special Theory of Relativity which addresses the motion in a straight line and in a uniform speed, Einstein started developing the General Theory of Relativity which was published in 1915. The General Theory provides a fundamental insight perception to accelerated movement, gravitation and cosmic geometry.

Gravitational forces have been correlated with accelerated movement by the classical laws of mechanics established by Newton.

Newton's second law of motion states that: Force = mass x acceleration (F = m * a)

Consider the case of your standing in a high speed elevator ascending from ground level to – say – level 500 in a high rise tower (figure 10-14).

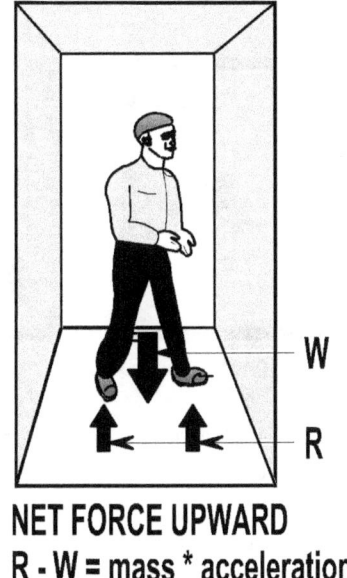

Figure 10-14: Forces experienced by a passenger in an

W

R

NET FORCE UPWARD
R - W = mass * acceleration

In the first few seconds of the trip, while the elevator will be increasing its speed gradually from 0 to its maximum speed; you feel its floor pressing upward against your leg and an upward reaction force is being transmitted through your body – e.g. a pressure exerted by your knee joint on your thigh and so on

This upward reaction force is what you perceive as an increase in the gravitational force you are used to experience.

According to Newton's second law of motion, **(R – W) = m * a**, or **R = W + m * a**, where R is that upward reaction exerted on you, W is your own weight in downward direction, m is your mass (often referred to wrongly as weight) and a is the acceleration you are moving at which is the same as the elevator's acceleration in this example.

While you are moving at an acceleration upward, reaction "R" which conveys the feel of gravitational force will be larger than your weight W by an amount = m*a, so you will be feeling an increase of gravity force. If the acceleration upward reaches 9.81 m/ sec2 (gravity acceleration) your feel of gravity will be doubled (right hand side in figure 10-15)

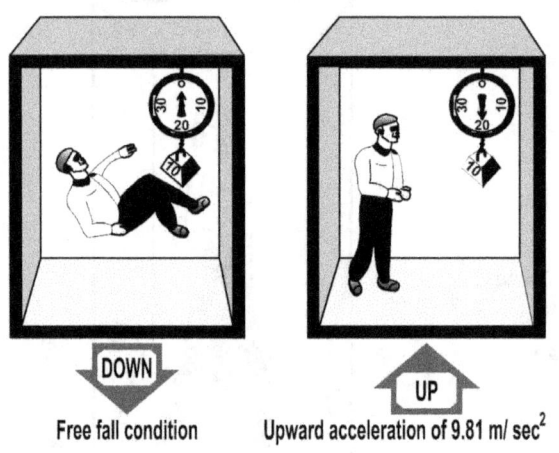

DOWN   UP

Free fall condition   Upward acceleration of 9.81 m/ sec^2

Figure 10-15

After that, when the elevator moves at a uniform speed;   a = 0 hence R = W, that is the upward reaction on your body will be equal to your weight, so your feel of gravity will be normalized.

Subsequently, and shortly before the elevator reaches its destination; it will decelerate (its velocity will gradually decrease till it reaches 0 which is equivalent to having a negative acceleration); so the magnitude of R will become less than your weight W, so you will be feeling a decrease in gravitational force. If such deceleration (negative acceleration) reaches 9.81 m/ sec^2, R will equal 0, so you will actually be in a free fall state (weightlessness) as if you were in deep space away from any gravitational field (the left hand side of figure 10-15)

It has therefore been established by the Newtonian physics that upward acceleration increases the effect of gravitational force, and downward acceleration counteracts the gravitational force and mitigates its effect.

EQUIVQLENCE PRINCIPLE

Einstein's Equivalence Principle states that a person in a freefalling elevator cannot tell whether he/she is at a state of free fall or in deep space environment; in absence of gravitational field like that one in the left-hand side of figure 10-15

Similarly, a person experiencing the effect of gravity in an enclosed space cannot tell whether he/ she is accelerating upward or he/ she is at rest in a gravitational field

GRAVITY

According to Newton; gravity attraction is caused by an invisible force that pulls two objects towards each other. If the mass of the two objects is **M1** and **M2** respectively and the distance between their respective centers of mass is **R**, the force of gravity **Fg** as established by Newton's law of gravitation equals

$Fg = C. M1.M2 / R^2$ where C is a constant.

Newton's explanation is that a gravitation force **Fg** of the magnitude computed by Newton's equation pulls the **sun** and the **earth** towards each other.

This gravitational force is neutralized by the centrifugal force F_{cent} resulting from the rotation of the earth around the sun

$F_{cent} = M1. V^2 / R,$ where M1 is the mass of the earth, **V** is the tangential speed of it and **R** is the radius of rotation (which is also the distance between their respective centers). Since the two forces **Fg** and F_{cent} are equal in magnitude and opposite in direction therefore

$$C. M1. M2 / R^2 = M1. V^2 / R$$

The General Relativity explains gravity differently. It states that: **"large masses cause the space-time to be curved and to wrap around these masses. Other objects will be forced to follow paths along the curved space-time that involve the least work"**. An artist's impression for the curvature of the space-time around massive objects is shown in figure 10-16

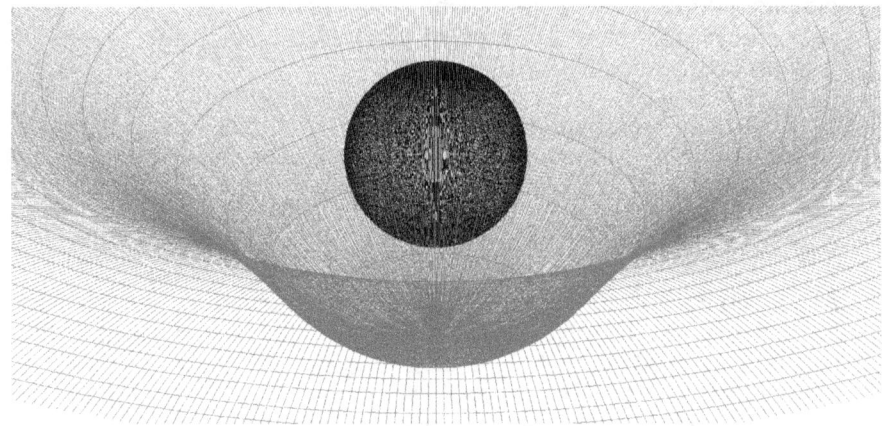

Figure 10-16: An artist's impression for space-time wrapping around a massive object

Bending the space-time as such is analogous to bending a sheet of thin plastic fabric held taut at its four points; by dropping on it a heavy iron ball.

The heavy ball will cause deflection to the fabric similar to that shown in figure 10-15. At this point; if you drop a smaller marble at the edge of the sheet it will roll down along the shortest path towards the ball. To any observer not seeing the fabric, the heavy ball will look like attracting the marble by an invisible force (Newtonian explanation).

An observer who can notice the groves created in the fabric by the heavy ball will realize that the marble finds its way through the shortest path along these groves

We have to pause here for a while to review the concept of shortest path through a curved space and the concept of curvature (how can it be detected)

If you want to travel through the shortest path from Paris to New York, you will not follow a route that connects the two cities literally by a straight line that penetrates the globe. You would rather follow an arc between the two points in on the earth surface. Unlike the case of two points in a plane in which the shortest path is the straight line joining them, the shortest path between two points on a spherical surface is an arc in a great circle (a circle that has the same diameter of the sphere). Shortest path in this case is called a "**geodesic line**".

The relativistic explanation of the rotation of the earth around the sun is that the sun causes the space-time continuum at its surroundings to wrap around it which forces the earth to move in an orbit that follows an elliptic geodesic line on the curved space-time.

This is illustrated in the artist's impression shown in figure 10-17

A question that may pop up in this context is "*how we can detect the curvature of space-time?*"

Let us first agree on the concept of curvature, and on a method to identify whether a surface (or a space) is curved. Amongst several of these methods; measuring the angles of a geodesic triangle stands as the most viable.

The sum of interior angles of a triangle drawn on an un-curved plane is 180° (figure 10-18). This is known as Euclidean triangle.

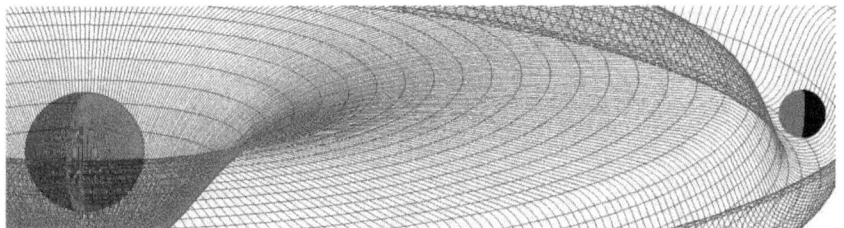

Figure 10-17: An artist's impression for space-time wrapping at vicinity of the sun forcing the Earth to follow a geodisic path around it

Non-Euclidean triangles drawn on a curved surface will have the sum of its interior angles exceeding **180°** if the curvature is positive. A curved surface is said to be having a positive curvature if the curve closes around a finite space like that of a spherical surface (figure 10-19).

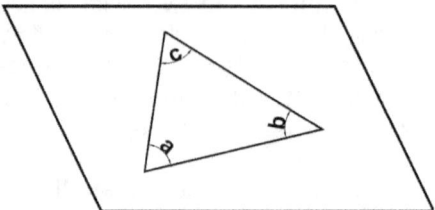

**Figure 10=18 Sum of interior angles in an Euclidean triangle
drawn on a plane is 180 degrees**

In case a curved surface tends to open at infinite distances such as the hyperbolic parabolic surface, it will be said to be having a negative curvature. Sum of interior angles in a geodesic triangle drawn on a negatively curved surface will be less than 180° (figure 10-20)

How can the theory of massive objects curving the space-time be tested?

The reader may suggest that triangular rope fencing is setup around a massive hill as shown in figure 10-21.

The hill is supposed to bend the space-time at its vicinity causing the sum of interior angles in the triangular fence to be larger or lesser than 180o depending on whether the curvature is positive or negative.

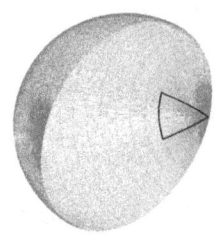

**Figure 10-19: Sum of interior
angles in a triangle drawn on
a positively curved surface is
greater than 180 degrees**

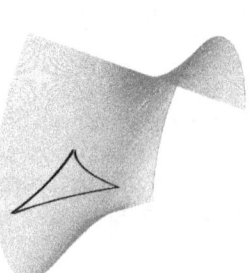

**Figure 10-20: Sum of interior
angles in a triangle drawn on a
negatively curved surface is less
than 180 degrees**

Unfortunately this kind of experiments does not work. It is like measuring the speed at which light travels for a distance of a few kilometers using stop watches to depict the starting and final times. Neither the hill nor even the whole Earth can create a curvature in the space-time which is large enough to be appreciable even by most precise instruments.

Probably we should use the Sun in lieu of the Earth to perform this space-time bending experiment. But how can we setup a triangular rope fence around the sun?

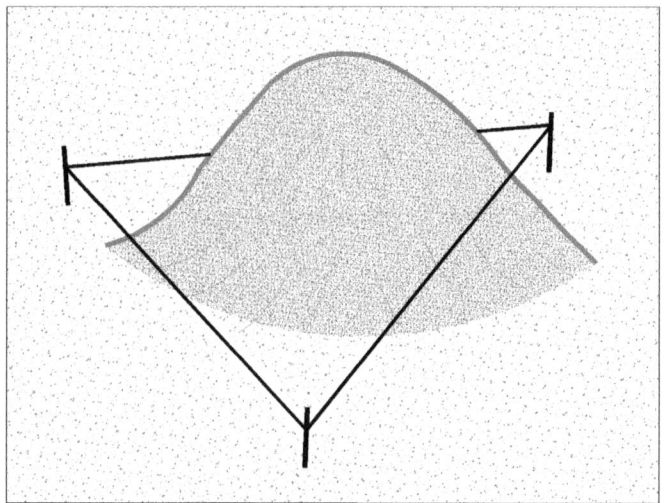

Figure10-21: Fencing a huge hill by a triangular rope to detect the space-time curvature at its vicinity

Einstein suggested that light beams are used instead. Why not? After all; light beams are also known to be following straight lines or to be precise, they follow geodesic lines of the shortest path. He further suggested that a starlight passing near the Sun in its way to Earth is to be examined to detect any deflection in its path due to curvature in the space-time caused by the Sun. Furthermore he predicted that the angular deflection of the starlight would be in the range of 1.75 seconds.

There is still a technical difficulty awaiting this experiment. The bright sunlight will make it almost impossible for any observer to see the starlight.

Again a solution proposed by Einstein is to perform the experiment during a total solar eclipse. In a solar eclipse the moon will be positioned between the Sun and the Earth and it will prevent the sunlight from reaching certain spots on Earth which will make stars visible even at daytime.

In the year 1919 an opportunity emerged! A total eclipse was due to take place on the 29th May and during the six minutes of its occurrence the Sun would be right in front of a cluster of bright stars.

A British astronomical expedition lead by the leading astrophysicist Arthur Eddington conducted the experiment at the Principe Islands at West Africa.

A schematic diagram for the experiment is shown in figure

10-22

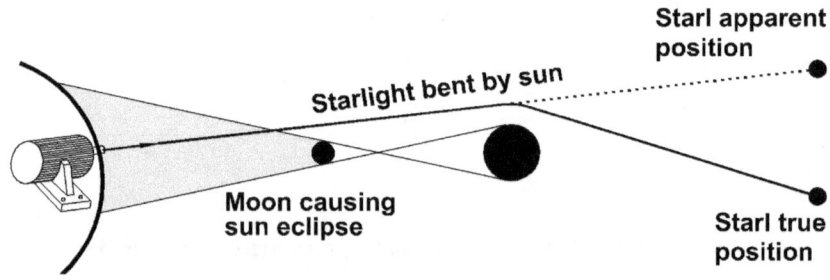

Figure 10-22: Measuring the deflection of a starlight as it passes through a space-time curvature created by the sun

The star angular position during the eclipse was taken and was compared with its angular position when the Sun was away from the path of the starlight. The angular difference between the star true position (when its light is not bent by the Sun) and its apparent position at the experiment was 1.61±0.30 seconds. This is pretty close to the 1.75 seconds predicted by Einstein.

The experiment constituted a strong confirmation of the General Relativity and was regarded as a great triumph of Einstein's relativity which was quite young and untested at time.

GENERAL RELATIVITY WRAPPED UP

According to the General Relativity theorem:

- **The effect of accelerated movement is not distinguishable from the effect being influenced by a magnetic field. A person experiencing the effect of gravity in an enclosed space - as that illustrated in figure 10-23 – cannot tell whether he is accelerating upward or is at rest in a gravitational field**

- **Mass causes the four dimensional space-time continuum at its neighborhood to be curved.**

- **Curved space-time will form the path – geodesic lines - for other masses to move around**

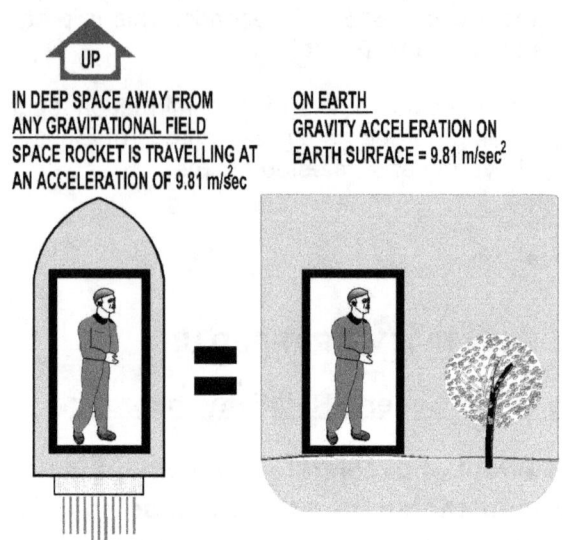

Figure 10-23: The effect of acceleration is equivalent to the effect of gravity

REFERENCES

- *Alexey Stakhov and Anna Sluchenkova, (..),Museum of Harmony and the Golden Section*

- *Amird D. Aczel, 1997, Fermat's Last Theorem, Dell Publishing*

- *Barbara Ryden, General Relativity – Key concept: General Relativity in one sentence, Accessed [Online] on August 14, 2015, Available from:* http://www.astronomy.ohiostate.edu/~ryden/ast162_6/notes24.html

- *David Acheson, 2010, 1089 + ALL THAT, Oxford University Press*

- *David M. Burton, 2011, The History of Mathematics – An Introduction, Seventh Edition, McGraw Hill*

- *Florian Cajori, 1909, A History of Mathematics, The Macmillan Company*

- *George Gamow, 1961, One Two Three Infinity, Dover Publications Inc*

- *Ian Stewart, 2014, Casebook of Mathematical Mysteries, Profile books*

- *John Stillwell, 2010, Mathematics and its history, Third edition, Springer*

- *Keith Devlin, 1998, The language of Mathematics, W.H. Freeman Henry Holt and Company*

- *Luke Hodgkin, 2005, A History of Mathematics From Mesopotamia to Modernity, Oxford University Press*

- *Ivars Peterson, 1988, The mathematical tourist, W. H. Freeman and Company*

- *Peter M. Higgins, 1998, Mathematics for the curious, Oxford University Press*

- *Sir Thomas Heath, 1981, A History of Greek Mathematics, Dover Publications Inc.*

- *The number mysteries, 2011, Marcus Du Sautoy,*

- *Victor J. Katz, 2009, A History of Mathematics, Third Edition, Addison-Wesley*

- *William Dunham, 1990, Journey through Genius, Penguin Books*

Index